# Advanced Litigation Support & Document Imaging

V. Mital (ed.)
*Reader for Lovell White Durrant, Brunel University*

*Contributors:*
R. G. Allison
N. Cameron
P. Deets
P. S. Eyres
H. Field
M. E. F. Fitzmaurice
M. C. Gowen
T. J. Heiden
J. M. Howie, Jr.
P. Illion
M. Macdonald
J. Mackintosh
J. B. Massopust
V. Mital
G. Pearson
R. Susskind
C. Thorne

Computer / Law Series 15

 Seminars

1995
Kluwer Law International
The Hague / Boston / London

A C.I.P. Catalogue record for this book is available from the Library of Congress

ISBN 90-411-0113-6

Published by Kluwer Law International,
P.O. Box 85889, 2508 CN The Hague, The Netherlands.

Sold and distributed in the U.S.A. and Canada
by Kluwer Law International,
675 Massachusetts Avenue, Cambridge, MA 02139, U.S.A.

In all other countries, sold and distributed
by Kluwer Law International,
P.O. Box 85889, 2508 CN The Hague, The Netherlands.

Cover design: Bert Arts

*Printed on acid-free paper*

# Contents

# 1.  Litigation Systems as Office Technology

*V. Mital*
*Brunel University*

When a case involves fifty or a hundred thousand documents, it is very difficult to manage the paper and produce necessary lists by using entirely manual     methods. Computerisation is the answer. Once the information about the documents is held in a computer database, a team of lawyers can go on to search for incriminating patterns, locate a letter that contradicts direct testimony given a day ago, find the smoking gun. So goes a powerful argument in favour of computerised litigation systems.

But few English firms have fifty thousand document cases, even fewer ones with a smoking gun waiting to be discovered. Save for the very largest firms, the above are hardly sufficient justifications of the cost-effectiveness of investment in litigation systems, for it is always possible to call in an experienced litigation management company should the need happen to arise. The fact that so many firms are still going ahead with pilots or operational systems is perhaps more indicative of the desire to be seen to be abreast of technology, rather than the calculated expectation of direct return.

In order to appeal to most medium-sized and regional firms, litigation systems must be justifiable for cases involving no more than a handful of lever arch files, such as could otherwise be handled with help from an assistant and a couple of articled clerks. There must be reasons to use these tools without having to ask the client for a contribution towards the capital cost. In fact, litigation systems need to be treatable as close as can be to ordinary office technology. There are some such systems in use today, and their number will increase. They happen to rely upon a newly popular technology, document imaging, but that alone is not sufficient to provide the benefits spoken of here. What matters more is that the designers and suppliers, and the users firms, have not limited themselves to the conventional, focused role of litigation systems, and have instead sought to facilitate the broad tasks performed by litigators in the usual course of their work.

*V. Mital (ed.), Advanced Litigation Support & Document Imaging, 1–8.*
© *1995 UNICOM Seminars. Printed in the Netherlands.*

## 1.1    THE CONVENTIONAL ROLE OF LITIGATION SYSTEMS

Lawyers in many American jurisdictions typically find themselves receiving in discovery, say, five times more paper than do English lawyers in equivalent cases. The practice of taking detailed and extensive depositions further means that there is a far greater need to produce lists of documents that relate to specific points. There may in addition be an exacerbating factor: the litigation team may operate from geographically distributed sites, requiring the same information to be housed at or accessed from multiple locations. The traditional way of dealing with the problem is to adopt a brute-force approach: documents are computerised at a very early stage in the proceedings, often before lawyers have had much of a look at the potentially relevant documents. The computer is used principally to create indices by which documents can be searched for from different perspectives. There are various, familiar ways of creating these indices:

- manual indexing that may be confined to recording bibliographic information, perhaps taking in information about persons, ships, buildings, etc. appearing in the text, sometimes extending to subjective analysis of the importance of the document to issues in contention

- automatic indexing of the original text of the documents, whether obtained by keying in the data by hand, from word processed files or through electronically reading the text printed on paper (OCR).

The aim is to try and shift as much of the burden of organising and handling information as possible away from qualified lawyers to less expensive paralegals or even coders without any specific qualification, other than literacy. This does create the problem of managing a large, non-professional workforce, which most law firms are not equipped to deal with. Typically, the problem is solved by the simple expedient of assigning the bulk of the work to external litigation agencies.

There is no denying the very significant advantages obtained by this approach in the appropriate case. For instance, substantial labour savings can be made because the computer can search very quickly and unerringly and can produce various selective lists of documents that satisfy certain criteria thought to be of importance. However, the benefits do not significantly include helping senior lawyers in tasks that involve detailed scrutiny and analysis of documents. A recent study carried out at Broad, Schultz, Larson & Wineberg, a Los Angeles firm, is instructive in this regard (Wallace, 1991).

In the case under study, seven oil companies had been charged with price-fixing and breach of contract in respect of purchases from a state-owned oil field. In the studied assignment, a government attorney was to identify all sales between the defendant companies over a given 31 day period. Three million pages had been collected over several years. The documents were filed in a library.

In the first scenario, a paper-based index was used. The attorney went through this index and identified and noted 326 possibly relevant documents. Once the librarian had retrieved them, the attorney went to the library and browsed through the selection. Having found 66 of the documents to be of particular importance, he requested copies for further perusal. On reviewing this last selection, the attorney found evidence of suspect transactions between one of the defendants and three as yet uncharged oil companies. By this time, the attorney had spent five days on the task. A further five days were spent by him in order to pin point the documents concerned with the newly identified transactions. During all this, the librarian spent a total of six days on physically retrieving, copying and refiling documents.

In the second scenario, a conventional LSS was available. As such, the attorney was able to search for the initial 326 documents in about half the time taken with a paper-based index. However, little else changed. Instead of a total of ten days, the attorney this time took eight. The librarian was still needed for six days: the computer gave access to an index, not to the documents themselves. The presence of the computer did not lead to a significant alteration in what the lawyer could delegate in respect of a task which required a high level of skill, namely, detailed analysis of documentary evidence.

Given the sizable document population, it is significant that there is such a small difference between the efforts needed from a litigator with and without a computer. The presence of the computer does not lead to a significant alteration in what a lawyer may delegate in respect of a task which requires a high level of skill. Such investment in technology does not significantly leverage the skills of the senior members of a law firm.

It is equally difficult to contend that a major, quantifiable contribution is made to the firm's profits. There is a reduction of costs, mainly in the relatively mechanical part of the work, but this is of a type that would result in a corresponding reduction in fee income. This is good for the client, and one who is aware of the implications would perhaps gravitate to a firm with computerised litigation support, or perhaps not. For other factors are still more important than such cost control and, in any case, an outside litigation support agency can do all the necessary work in the occasional case that cannot be handled without computer support.

In other words, it is quite possible to live without such computerised litigation support, and most law firms have proved that to be so. The question then is, has the situation changed? To answer it we should take a brief look at newer kinds of systems that rely on document imaging and, sometimes, also on other relatively new technologies. It must, of course, be emphasised that mere addition of document imaging to a litigation system in the conventional mould would not cause the question necessarily to be answered in the affirmative, though the balance of arguments may indeed be shifted somewhat.


## 1.2     DOCUMENT IMAGING

Basically, in the context of litigation support, document imaging involves taking a snapshot of a piece of paper, storing it in an electronic form, displaying it on the screen and producing a paper print when needed. The paper original - a paper copy from which the electronic image was taken may be thought of as the original for this purpose - may be retained for production before a court; the litigation team need not resort to the paper original until such a time. The implications are quite obvious: a number of persons can simultaneously have ready access to the same document; a case library need never be incomplete, as it would be from time to time if documents were physically removed for copying or referral, and so on.

The technology has been around for many years, but was until the late 'Eighties too expensive to be used in litigation support. The capital cost is still quite high. In the case of hardware it is now falling only moderately. The software component, on the other hand, is due to come down dramatically in price as developers begin to take the opportunity to generalize and market applications which were originally developed for custom clients. As for operational costs, a full account, including details such as how much it costs to have a page imaged or converted to text, is available elsewhere (Massopust, 1992).


## 1.3     ADDING DOCUMENT IMAGING TO CONVENTIONAL LITIGATION SYSTEMS

The approach to litigation support described above can be augmented with document imaging. Whereas previously the system held only some indexing information about a document, now it can store an image of the document for the user to view or print from as required. There is no longer the need to go to a document library to browse through documents or to request someone else for copies of special interest. The savings in effort can be remarkable. In fact, this can be illustrated by further results from the above-mentioned study at Broad, Schultz, Larson & Wineberg.

The same assignment as mentioned before was carried out using an image-capable LSS. This time, the attorney required just 4 days to complete the tasks. Further, this time there was no need to involve the librarian, saving 6 days of work from that source.

The table below shows how the work changed when an imaging-capable system was available, in comparison to that when an ordinary LSS was used.

| | COMPUTERISED LITIGATION SUPPORT WITHOUT IMAGING | | COMPUTERISED LITIGATION SUPPORT WITH IMAGING | |
|---|---|---|---|---|
| ACTOR | ACTIONS | TIME (days) | ACTIONS | TIME (days) |
| Attorney | Searching for transactions of interest; identifying 326 possibly relevant documents | 1 | Searching for transactions of interest; identifying 326 possibly relevant documents; viewing the images of these documents; printing copies of 66 of them | 1 |
| Librarian | Retrieving 326 documents | 2 | - | - |
| | | | | - |
| Attorney | Examining the 326 documents requesting copies of 66 relevant ones. | 2 | - | |
| Librarian | Copying 66, refiling 326 documents | 1 | - | - |
| | | | | 1 |
| Attorney | Reviewing 66 documents; finding that suspect sales also occurred between one of the defendants and three uncharged companies | 1 | Reviewing 66 documents; finding that suspect sales also occurred between one of the defendants and three uncharged companies. | |
| Attorney | Go through above three steps | 4 | Go through above two steps | 2 |
| Librarian | Go through above two steps | 3 | - | - |

*Table 1    Change in work with imaging*

These results are much more encouraging. There is now more time to spend on improving the quality and thoroughness of the case analysis. But things could be better: for instance, it is still not possible to use a junior fee-earner to do part of something which previously required a partner.

There is clearly a great improvement in efficiency and reduction in total costs, but it is not certain that the profits have gone up in step with the magnitude of investment. For a very large case the client may well agree to fund much of the capital investment. It is more difficult to see where the returns would be coming from in the run of the mill case.

As firms and developers have learnt lessons from early forays into document imaging, they have realised that it is not sufficient to continue to have technology tagged on as an occasional digression from the normal work practices of litigators. Newer systems aim to harness the power of advanced technology and to integrate it into the normal, everyday work practices. They go well beyond holding computerised indices to paper documents and instead seek to facilitate working with those documents in a rich manner. Some of the important contributions made by such systems are briefly discussed next.

## 1.4     INCREASING THE MARGINS ON COPYING

Document imaging can be thought of as deferred photocopying. Unlike in the case of a normal photocopier, it is not necessary that the paper copy should be produced immediately after a snapshot has been taken of a piece of paper. As many copies of documents as needed, in any desired order or combination, can be produced. Except for the effort required when electronic images are being taken, the manual effort involved in handling paper and attached chores such as destapling and reassembly of documents is virtually eliminated.

The labour savings to be obtained thus have been recognised for some time. However, until earlier this year, there was present a counter balancing factor: the cost of consumables in producing prints from images was around three times higher than the corresponding per page cost for a heavy duty photocopier. Moreover, except in the case of machines costing over £100,000, image printers were of the small, desktop variety and it was not possible to get large paper magazines and sorters. As the market has developed, this situation has changed. All in all, the cost savings now available can be well over four pence per page. As a rule of thumb, the core documents will need to be printed at least seven times each during the course of the litigation. In a case with five thousand documents (another rule tells us that each document has 3.4 pages), this gives an added margin of more than £4,500 under this head alone. Until such time that most firms have imaging technology, it is difficult to see why this should not accrue wholly towards increased profits.

Thinking of the litigation support system also as a substitute for the photocopying department requires some forethought in the way documents are coded and analysed. For instance, a document may be marked for later inclusion in a number of different bundles, and some of this marking may be carried out when lawyers are in the process of reviewing individual documents of interest.

## 1.5 MORE EFFECTIVE TEAM WORK

Many senior fee earners delegate to their juniors the task of analysing important documents in the first instance and making notes about them. It gives the senior lawyers greater assurance in the quality of work done by the latter to know that, should they have a hunch that some comment or conclusion is erroneous, they can call up the original on the screen in an instant and review the matter personally.

The notes themselves are best made *in situ*, i.e. annotations are most informative, and the ambiguity of interpretation least, when they are placed adjacent to the appropriate paragraph of the letter or feature of the drawing. A number of other persons, including expert witnesses, may end up commenting on some of the documents. Imaging technology allows multiple layers of annotations; it is possible for the viewer selectively to obscure some or all of the notes arising from a particular source. The level of functionality available is far higher than what is achievable by annotating photocopies.

Essentially, the litigation system acts as a fast open channel of communication between the team members. It is possible for one lawyer to send his view of some evidence to another or seek opinion on a problematical set of documents, all through the system. The process is set up so that a lawyer need not know anything about computer communications and need only drag an icon on his screen representing some documents on to another representing the relevant addressees. It is possible to assist the process of managing the wider litigation activity in various ways; for instance, the system may be programmed so that where two persons are to review a document, and one is not able to do so within a given time, the other is automatically informed.

## 1.6 USING THE VALUABLE INFORMATION HELD IN MANUAL FILING STRUCTURES

When we file documents by hand we do not need to record in writing everything about them. Much is gathered from the context (Mital and Johnson, 1992). That one document is placed in a folder several places behind another gives some impression of the relevant chronology. The fact that three documents are clipped together under one covering letter is useful information; the order of the former documents is important too, because one of them may be an attachment of another. In a conventional litigation system such visual cues are lost, because all the paper is reduced to an amorphous mass. Consequently, the indices which are created need often to be more detailed than would otherwise be required, the cost of producing them is correspondingly higher. This is not to say that filing paper in the traditional manner is the answer, for that is subject to much inflexibility. It is not easy to change the order of documents or to look at cross-sections of information held in paper files. It is just that there is a cost if we lose the rich information contained in the paper received by a law firm.

Firstly, greater effort has to go into indexing the documents up front. Secondly, and more critically, lawyers have great difficulty in making their way through information that has lost familiar structures and require much additional training. Graphical user interfaces and information visualisation techniques being introduced in the latest systems go some way towards countering these difficulties. Essentially, the intention is to allow users to cross-relate directly between what is found in a well-organised paper-based office and what the computer holds. In fact, the early 'Eighties vision of the paperless office has been virtually abandoned, for paper is often the best means of appreciating and working with information. Instead, the shift is towards the paper-where-needed culture.

## 1.7    CONCLUSIONS

Litigation systems targetted at the largest of cases are usually of not much help in the more common types of litigation and it is difficult to obtain a proper return on investment in such systems. In fact, there is some sort of an analogy to be made with the adage that difficult cases make bad law. Investment in more flexible, newer types of litigation systems can be treated on par with other office technology. A number of tangible benefits, of which a few are described here, are obtainable in the form of increased effectiveness of substantive practice tasks and a contribution to profits through greater margins in the provision of ancillary services.

## REFERENCES

Massopust, J.B. (1992) Understanding the Detailed Economics of Imaging, *Proceedings of the Document Imaging in the Law Office Conference*, Brunel CCLF and Unicom Seminars, Uxbridge.

Mital, V. and Johnson, L. (1992) *Advanced Information Systems for Lawyers*, Chapman and Hall, London.

Wallace, S. (1991) The Image-Enabled Office, *Office Computing Report*, **15**(1).

# 2. Document Image Processing

*R G Allison*
*Allison Consultancy*

## 2.1 WHAT'S THE PROBLEM?

All businesses handle large amounts of paper. In some cases this presents no significant management problem, but if any of the following are true then a problem probably exists:

- you have to respond to a wide variety of documents that originate outside your organisation (and hence outside your control)

- the physical speed at which the paper moves around your business determines (and limits) the pace and response of your operation

- access to your documentation is too time consuming

- your documents have a high intrinsic value so their loss is expensive, or exposes the business to risk of prosecution or restriction in operations

- document retrieval, transmission or handling delays limit your business opportunities

- there is intensive, structured manual document processing

- the size of your filing registry is growing and extra on-floor space would be very expensive, or simply not available.

- you have complex retrieval requirements, often involving searching through the contents of documents rather than simple file reference.

*V. Mital (ed.), Advanced Litigation Support & Document Imaging, 9–14.*
© *1995 UNICOM Seminars. Printed in the Netherlands.*

## 16.2    WHAT'S THE SOLUTION?

If any of the above scenarios occurs in your business, you may benefit from a technique called Document Image Processing (DIP). With DIP, pages of information are scanned into the system using a machine similar to a photocopier. The scanned pages (images') are stored in the computer, and can be displayed on screen or printed out. DIP can handle any information, including printed text, diagrams, handwriting, logos, signatures etc. If you can photocopy it, you can scan it. This makes for very flexible systems.

## 2.3    WHAT ARE THE BENEFITS?

| | |
|---|---|
| Customer service | Royal Life reduced their response time to customers from 24 days to 4. NatWest credit card reduced theirs from 9 days to 1. |
| New product/ project implementation | most new products and projects involve multi-disciplinary teams, geographically dispersed. DIP can greatly improve the flow of information and hence allow the product to be brought to market or the project to be brought on stream earlier. |
| Product quality | TQM places great demands on all parts of an organisation to communicate effectively and rapidly. DIP can enable this. |
| Control | manual procedures do not provide much information that is useful to aid control, most is after-the-fact monitoring which only allows for fire-fighting. DIP allows pre-emptive control of business procedures |
| Productivity | In structured working environments ('back office') the handling of documents in electronic form can yield very significant improvements in productivity. Citibank increased productivity by 35%. Norway Central Bank have saved 33% in their data input (=100 people). |

## 2.4    HOW MUCH DOES IT COST?

A small system that would serve a small number of users can cost under £30K. A larger system which could serve up to 100 or more users and have several million pages of information on-line would cost £100-250K for the central 'server' Individual additional users cost between £200 for adding image capability to a Windows based

PC to £1-2K for adding large screens and drivers. The former would be suitable for casual users, the latter for heavy users. A significant extra cost can be incurred in loading a large backlog of existing documents into the system - allow 10-15p/page.


## 2.5   WHAT'S THE ALTERNATIVE?

The following techniques can be used in some cases, either instead of or complementing DIP. However, DIP is like death and taxes - it's inevitable. Your business will be using image and related technologies at some point, it is merely a question of time.

| | |
|---|---|
| Do nothing | cosy, but what about the competition? Full-scale introduction of a major new technology like imaging takes 2 years from initial idea. This is too long to allow 'catch-up' after the competition has made it's move. |
| Review of Operations | sometimes a review of existing operations can lead to savings - bureaucracy is like ivy, it slowly spreads until it chokes the entire edifice. Occasional pruning is essential. |
| Conventional IS Systems | the use of electronic mail can greatly improve internal information flows, although it will not tackle external paper documents. Document index databases can aid searching. |
| Optical Character Recognition (OCR) | will read the text of printed documents so it can be searched and manipulated by conventional systems. For high quality, unstructured text documents this can be very useful. |
| Electronic Data Interchange (EDI) | Here all partners in a business relationship agree a standard form for electronic transfer of coded business information. This offers major benefits where such an infrastructure has been built up. |


## 2.6   RELATIVE MERITS OF PAPER, MICROFILM AND IMAGING

The following table shows the relative merits of paper, microfilm and imaging. These apply in all businesses, but of course the benefits are of differing relevance in each area. It can be seen that imaging offers greater benefits than microfilm in almost all areas, but at a significantly higher cost. There is no reason why a hybrid solution cannot be adopted which uses an appropriate mix of all three media.

| Attribute | Paper | Microfilm | DIP |
|---|---|---|---|
| Space required (more 33 = less space) | 3 | 3333 | 3333 |
| Flexibility of retrieval | 3 | 33 | 3333 |
| Multiple access to information | 6 | 33 | 3333 |
| Remote access to information | 6 | 6 | 333 |
| Ease of systemisation of processes | 3 | 3 | 3333 |
| Speed of access | 3 | 33 | 3333 |
| Speed of updating files | 33 | 3 | 3333 |
| Ease of use (once accessed) | 333 | 33 | 33 |
| Legal Admissibility | 3333 | 333§ | 6 |
| Cost (more 33 = lower cost) | 3 | 333 | 33 |

§ - microfilm legality is good if properly certified, but microfiche legality is poorer

| Benefit | Paper | Microfilm | DIP |
|---|---|---|---|
| Improved Customer Service | 3 | 33 | 3333 |
| Improved Productivity | 3 | 33 | 3333 |
| Integration with Existing Systems | 3 | 3 | 333 |
| Space saving | 6 | 333 | 333 |
| Flexible siting of staff nearer to customers | 6 | 6 | 333 |
| Future-proofing | 3 | 3 | 333 |

The benefits outlined above are:

### 2.6.1   Improved Customer Service

Consider the scenario where a customer telephones in with a query. With paper or microfilm systems it is usually necessary for the clerk to leave his desk to search for the customer's file. If the customer has a transaction in progress, it is likely that another clerk will have pulled the file and hence it may not be available. This means that the clerk has to tell the customer he will ring back. This often then leads to 'telephone tag' With imaging, all information is available on the clerk's screen. The typical response time would be 5-10 seconds with a worst case response of 15-20 seconds. This ensures that the clerk can handle the customer's enquiry on the phone immediately. This improves the customer's opinion of the unit's efficiency as his details are always to hand, and the clerks are fully aware of his case.

### 2.6.2  Improved Productivity

Where the work of an area is well structured with information taking a fairly well predefined route, imaging can provide very high productivity savings by the use of 'workflow' techniques. For example, an invoice coming into the unit would be scanned in and relevant key information keyed in. The system would then automatically retrieve the necessary supporting documentation, including details from the mainframe IT systems. This information would be bundled together as an 'electronic dossier' and would be routed to the appropriate person for action based on the details of the invoice - its value, the business concerned etc. When the clerk asked for the next invoice, he would be presented with all the relevant information together. In highly structured environments like these productivity gains of 40-50% can be realised.

### 2.6.3  Integration with Existing Systems

Any documentation handling system must integrate with the existing IT infrastructure in order to gain maximum leverage. This is difficult with paper and microfilm as they are physical rather than electronic methods. Imaging can be closely integrated, although this does take some effort.

### 2.6.4  Space Saving

Both microfilm and imaging can provide dramatic savings in space and weight. One 12" optical disc can hold 120,000 pages of A4 images. The optical disc weighs about 1kg and takes 0.5" of shelf space. The equivalent paper weighs 600kg and takes 60 feet of shelf space. Optical discs can be held in a 'jukebox' which will load the required platter automatically. A high capacity jukebox is about the size of a desk and can hold about 6.5 million page images, which is equivalent to 3,300 ft of shelf storage.

### 2.6.5  Flexible Siting of Staff

It is perfectly possible to access image information over a network. This makes it possible to have a more flexible policy for siting of staff. In particular, it is possible to move staff closer to the customers and yet retain full access to information. It is also possible to allow customers to have limited access to the system to make their own enquiries, thus reducing staff requirements.

### 2.6.6    Future Proofing

There is no doubt that all businesses will be using imaging systems at some point in the future. The only question is when is the optimum time for any one unit to make the transition. It is still important, however, for businesses to consider standards in order to ensure that they do not end up with an obsolete system.

## 2.7    WHAT NEXT?

- define the business case — the sequence of events should be:

    i   review your records to see if they truly support your business
    ii  review your filing and retrieval requirements to ensure you can access your information
    iii provide an effective mechanism for handling your documents — this <u>may</u> be via DIP, but could also be via more efficient paper handling, microfilm or OCR.

- have a joint end-user/IS Dept/Records Management team approach.

- seek advice as DIP skills are still rare.

- set up a pilot system — this allows you to tackle the problem in bite sized chunks, but make sure you have a large enough scope to achieve critical mass.

# 3. The Litigator's Perspective on Being Supported

*Thomas J. Heiden*
*Miller, Canfield, Paddock & Stone*

In both the state and federal courts, there have been civil cases involving "universes" of documents for several decades. The American courts have permitted general document discovery (production, inspection, copying) for years. At about the same time, the copy machine expanded the number of documents available for production exponentially.

In the 1950's and 1960's, the massive document discovery cases were largely cases arising under the anti-trust laws. In the 1970's, large construction projects (often multi-billion dollar nuclear powered electric generating plants) spawned mega-cases, some involving tens of millions of pieces of paper.

Other types of cases have become equally document intensive - product liability pattern litigation, bond and securities litigation, lawsuits over the Savings and Loan bailout program.

As court rules permitting general document discovery were combined with the power of the copy machine, trial lawyers were confronted with a new problem — a document management problem. They also were presented a new opportunity — a chance to comb millions of pieces of paper to find the "smoking gun" — the one piece of evidence that will turn an entire case. And find it they sometimes did. American legal folklore is sprinkled with stories of that one "hot document" — the one that proved a plaintiff's claim, made a lie of a party's entire case, or drove an enormous settlement. Some of the folklore are apocryphal; some are true.

But finding that one "smoking gun" in a ten million page haystack while also keeping track of the hundreds of important documents, the thousands of relevant documents and the nine and one-half million other pieces of paper required an assist from technology. Technology provided that document management assistance in the form of computerized litigation support.

Both building and using early systems were cumbersome and difficult.

*V. Mital (ed.), Advanced Litigation Support & Document Imaging, 15–20.*
© *1995 UNICOM Seminars. Printed in the Netherlands.*

Document collections were often reviewed several different times. Often the inspection was done by lawyers or expert witnesses. The lawyers and experts filled out long coding forms by hand. Those forms were separately keyed and loaded into mainframe computers. Access to the data was limited — sometimes to business hours, usually by narrow searching capability, always to a fixed office computer console. Most early systems could not be searched by the lawyers at all; rather, the law firm phoned a "batch" of information requests to the vendor who would gather the information overnight and deliver a print out the next day.

In American civil litigation today, litigation support systems are neither experimental nor trendy. They are tried and proven. They are simply an accepted part of big case management.

## 3.1    DO YOU NEED COMPUTER ASSISTED DOCUMENT MANAGEMENT?

In a complicated case, the trial lawyer needs lots of support. Lawsuits always involve legal research and fact development. Today's multi-media litigation often involves expert witnesses; computer animated graphics; jury science; and seemingly endless document review. Today's trial lawyer needs all that support in addition to the time-honored skills of advocacy. The trial lawyer does **not** need a mysterious computer system which the trial lawyer does not understand and does not know how to use.

Lawyers ought first to decide whether they need a computer-assisted information storage and retrieval system at all. It is really nothing more than an automated filing system — whose only purpose is to help the trial attorney win the case — by settlement, verdict or judgment. An old fashioned manual system may do just fine. If you can do it manually, do it manually. If you can remember it, then just remember it.

There is no arithmetic formula for deciding whether a litigation support system is needed. Consider one practical test: are the lawyers spending their time reviewing historical documents instead of doing the lawyering they were engaged to do. If "yes", the case probably needs computer-assisted document management.

If a computer-assisted system is required, the trial lawyer ought to participate directly in deciding what type of system. There are many systems and many vendors. Most of what they involve is foreign and uninteresting to the trial lawyer; the entire design and selection process provides an obvious invitation for delegation by the busy trial lawyer. Do not accept the invitation entirely — at least choose the criteria, the capabilities you want in your system. To decide think of, see yourself **using the information** on the system, using it in and for the lawsuit.

## 3.2    DEALING WITH THE IN-HOUSE LITIGATION DEPARTMENT

Some companies or clients have been through complex litigation before. They may know exactly what documents they have and how they are organized. They may have a system in place. Some clients are wholly uninitiated and will have no idea of what they have or what needs to be done. Others may think they have all the answers and will expect you to accept that.

Any complicated endeavor needs a general. Generals need authority, logistical support and armies. Don't let the in-house people superimpose their system on you. Remember, your goal is to win this suit. Cutting corners, using something simply because it was used before, telling in-house people what they want to hear — are not among your goals.

Rather, in deciding what sort of system is required, remember what is important. What kind of case is it? What court is it in? Who are the adversaries? What is the discovery schedule? What is the likely trial date? How many documents are there? Where are they? Are they your client's or their's? What kind of documents are they?

In deciding which vendor, ask a similar set of questions. Is the vendor asking the same questions you are? Has this vendor been to trial before? What do the people with whom the vendor has worked think of its people and its system? Make your decisions based on those criteria.

The trial lawyer would never want someone else to select his expert witness or pick his jury. In a complicated case, the litigation support system may be the most important support tool — from discovery through trial. Listen to the experts; but make your own decision on what you believe is best for you and your case.

## 3.3    HOW THE LAWYER CAN MANAGE THE OUTSIDE HELP

If the trial lawyer has invested his time up front, if he has participated in the selection of the system, then he is already managing the outside help.

There are, however, more decisions to make. Which documents are going to be included on the system? What information will be captured and how? What sort of coding? Objective coding, subjective or taxonomic coding, fact intensive issue coding? Decisions on optical imaging, full versus partial text, enhancing. Decisions on what to include in addition to the historical discovery documents — pleadings, hearings, depositions, trial testimony, briefs.

Set some basic standards for making those decisions and enforce them. As you make those decisions, imagine yourself using the system — for motion practice, at depositions, for settlement negotiations, in the courtroom at trial. What sorts of standards ought to be set and enforced? The most fundamental benchmarks. Cost. Get a cost estimate. Use it to measure the discovery effort, your vendor and its system. Schedule. Set a realistic schedule for each phase. Hold yourself and your vendor to that schedule. Deviate from the schedule only if the Court or some fundamental change in the size and shape of the suit directs a change.

## 3.4    HOW TO BE CONFIDENT OF THE INTEGRITY OF THE DOCUMENT BASE

To assure confidence in the integrity of the data base, use and remember two words: reliability and usability.

### 3.4.1  Reliability

There are some bad stories from the evolution of litigation support systems. Companies, who sold expertise and experience which they did not have. Coding shops that had low accuracy. Systems that were so unfriendly that trial lawyers despaired of actually using them. Efforts which fell so far off cost and schedule projections to rival defense department procurement standards. That does not happen to the good companies and the good systems and there are plenty of both.

Lawsuits and trials are not perfect. The data base need not be perfect. But, with the proper controls, the litigation support system can be close enough to perfect for the trial team to rely on it with confidence.

Insist upon demonstrated quality control from vendor. Do some spot check quality control of your own. Test the system - its objective and subjective capacity, the accuracy of its response. Do some of that early in the development of the system — early enough that there is still room and time for adjustments and refinements to the system.

### 3.4.2  Usability

The litigation support system needs to be something that the trial lawyer can and will use — by herself, at night, far from home, during trial - actually use in the most hostile litigation environment.

In the vendor and system selection process, look for companies that understand both computer technology and lawsuits. Look for those who have married a knowledge of hardware and software to actual trial experience.

The latest bells and whistles of computer science are usually less important than a commitment to the professional service of lawyers, lawsuits and trials. The system developed by people who provide state-of-the-art technology combined with an intimate knowledge of trials is the system you are most likely to actually **use**.

## 3.5    WHAT CAN GO WRONG AND HOW TO RECOVER

Lawsuits are dynamic; lawsuits change size and shape; parties come and go; issues come and go. Those dynamic changes affect all facets of pre-trial strategy — motion practice, fact development, expert witnesses. They also affect document discovery — whether manual or computerized.

1. Don't issues and parties change?

   Of course they do. Do not organize a litigation support system by party or by issue. Emphasize specific fact categories for all taxonomic or issue coding.

2. What if schedules change?

   Sometimes parties change schedules; often courts change them. Organize your effort to the shortest realistic schedule. Then hold your people to that schedule.

3. Should all documents be treated the same?

   Probably not. For instance, historical discovery documents can be objectively and subjectively coded, with the full text image on optical disk. The court record — depositions and trial transcripts - can usually be ordered on disk and loaded full-text onto the same system.

4. How can we trust issue coding to non-lawyers?

   Do not have lawyers do the coding work themselves. Their efforts are not in a controlled document management environment; often it is not methodical. Lawyers are trained and hired to do lawyering. Often they get bored with repetitive coding work. And lawyers are far too expensive. Screeners and coders, if smart, will pick up things on their own. Their work is controlled, supervised and methodical.

5. What if we don't find the "smoking gun"?

   There may be no "smoking gun" to find. 99.9% of the work with the litigation support system will not be "smoking gun" work. Rather it will involve doing the drudgery — preparing for depositions — for motions — to rebut this point — to answer that contention — to explain something else. That work will be the real test of your team and your system. That work will develop the theme or story for the presentation of the case — methodically developing, adding, refining, checking the "themes", the "story". It is with that enormous — but not glamorous — effort that most cases are won or lost.

6.  Will we have spent a lot of money for something we won't use?

At trial, not only need the right fact be found but it must be found in time. These systems work hard — I have seen them used and used them — during long trials, several thousand miles from the location of the lawyers' offices and the documents themselves, seven days a week, almost literally twenty-four hours a day.

# 4. Computers in Support of Litigation

*Richard Susskind*
*Masons*

It is hard to imagine a large law firm of today operating without information technology (IT). Word processing, accounting systems, electronic mail and many more administrative and management applications are widely used. In contrast, lawyers themselves, in their advisory capacity, are far from dependent on technology. It requires little vision to imagine lawyering without IT: take technology away from most lawyers and there would be no perceptible change to their daily lives.

But lawyers — as distinct from law firms — cannot remain impervious to technology for much longer. Litigators, in particular, are poised to undergo their own computer revolution. The term that is widely used here is "litigation support", one branch of which involves systems that help with the management and control of large quantities of documents. The potential of this technology is particularly clear in complex technical cases, such as construction or computer disputes, where the party that has mastery of the documents can enjoy a clear strategic advantage over others.

Three techniques currently dominate litigation support. One approach is to compile a computerised index of all documents relating to a case. Each document is represented in a database as a collection of objective features (eg, date of document, author, recipient) as well as subjective features, requiring lawyers' classifications (such as whether a document is privileged or prejudicial to the client's case). Once set up, such a system can sort all documents, for example, in date order or by author's name. Additionally, the system can search and produce lists of documents sharing particular features: for instance, a list of all privileged documents written by Company X to Mr. Y between two specified dates.

Another technique is to build an information retrieval system that holds not an index but the full text of a collection of papers. This enables lawyers to search quickly and easily within the entire text of documents for the occurrence of single words (e.g. names of individuals, companies, places, or terms such as "warranty" or "delay") or for words in combination (e.g. the name of a company within a specified number of words of the name of an individual or a phrase such as "defective software").

*V. Mital (ed.), Advanced Litigation Support & Document Imaging, 21–24.*
© 1995 *UNICOM Seminars. Printed in the Netherlands.*

A third approach to litigation support uses imaging technology. This process can be likened to taking photographs of individual documents and so this technology can cope well with non-textual materials such as drawings and handwriting. Users of systems that hold images cannot search for individual words within the imaged documents (the text is not in machine readable form). Rather, they can view these images as if perusing microfiche on a computer screen. A database containing only images has two main uses: first, in overcoming problems of physical space and storage; and, second, in reducing the amount of paperwork that needs to be handled in the courtroom by judges and juries.

The real benefits of litigation support will come with a combination of these three techniques. On one particularly promising model, the lawyers's first exposure to a set of documents will be through a litigation workstation with a large television-sized monitor. This system will contain the images of all the documents, each objectively indexed; and the lawyer will read through the images of the documents on screen, adding subjective commentary into the index and selecting (by a pointing device) portions of text to be converted from image into searchable form (using optical or intelligent character recognition techniques).

The end result will be a sophisticated index, a database of searchable extracts, as well as the images themselves. Such a system would be invaluable, for lawyers and clients, both in preparation for trial and in the courtroom itself.

Those who consider litigation as too confrontational, costly and time consuming often find in litigation support the makings of a panacea; a promising source of enhanced productivity, quality and efficiency and in turn the means by which disputes might be pre-empted, settled earlier or resolved at lower cost and greater speed.

So where's the rub? Why are all lawyers not using litigation support technologies?

Lawyers have a range of misgivings, most of which are rooted in their incomplete picture of what can actually be achieved. Lawyers' preference for secretive exploitation of IT does not help here (see *Financial Times*, Legal Column, 22/10/90); nor does their refusal to work together in pre-competitive, collaborative spirit in establishing standards and settling on compatible systems.

A further obstacle is the widely held belief that it can rarely be cost effective for the purposes of litigation to transfer documents from paper into machines. Historically, this has been true, but rapid advances in optical and intelligent character recognition technologies, together with the emergence (as in the U.S.) of external bureau services devoted to indexing, data entry and imaging, combine to suggest that this obstacle can now be negotiated. Moreover, as clients themselves increasingly use computer-based document management systems for their own administrative purposes, it will be a lesser task to convert from these systems to litigation support systems than the current challenge of moving from paper to system. In promoting this synergy between document management systems and litigation support systems, the more proactive lawyers in the profession should actually be advising on the potential compatibility of these systems.

One genuine problem is the unfortunate uncertainty over a crucial issue - costs. There has been no decision on whether the costs of setting up and running litigation support systems can be recovered by a successful party in litigation from the unsuccessful party. This is a matter being pursued by the Society for Computers and Law, as it is unacceptable for solicitors to be unsure when asked to advise their clients on the recoverability of development and staffing costs incurred in litigation support. A further, related anomaly arises from the general requirement that lawyers charge on the basis of the hours spent on a task; a phenomenon which could, in principle, reward the inefficient firm and penalise the well run practice. When litigation support systems are used, and time is saved, total time spent becomes a less reliable indicator of the value of a service. Cynics have suggested that lawyers will be reluctant to become too efficient with technology until there is a move beyond the billable hour. It would be sad if lawyers' uptake of litigation support technology was inhibited by inapposite billing and cost recovery practices.

Yet none of the perceived problems seems to be insurmountable, as the more progressive litigators are showing by their successful deployment of the new technologies. Indeed, ignoring litigation support is fast receding as a commercial option: as barristers with positive experience come to expect solictors to use computers; as courts encourage and require parties to employ database technology; and as clients realise that a higher quality, lower cost, wider ranging service is available from hi-tech solicitors, legal luddites will soon struggle for their day in court.

For clients, these developments raise challenging questions about the suitability of the lawyers they instruct. A further criterion in selecting legal advisers now emerges, relating to the extent to which lawyers have appropriate technology skills and support. If in major cases of the future, all parties have the documents held in litigation support systems (loaded perhaps by some external bureau), a key point of differentiation amongst practices will be law firms' relative proficiency in exploiting the data in these systems. Are the lawyers adequately trained in advanced searching techniques? What practical experience and track record do they have with litigation support? Do they have permanent, first rate support staff? Are they using advanced techniques, such as conceptual searching, intelligent "front-ends" and hypertext to enhance the basic systems? Are they capable of advising proactively on versatile document management systems? Do they understand the complex legal questions, regarding issues such as admissibility and authentication of evidence, that litigation support systems raise?    These are questions clients should shortly be asking.

A question today for all clients is whether their current lawyers are investing sufficiently in IT in preparation for the central role it is destined to play. The stage is set for major change in the world of litigation: in five years, complex, large scale litigation in this country will invariably be supported by IT; while by the turn of the century, litigation without IT will be virtually unimaginable.

# 5. The Application of EIM/OCR Technology to Litigation Case Management

*Phyllis V. Deets*
*Gowen Deets*

This paper discusses the application of electronic imaging technology to the field of computerized litigation support. Whether the particular legal dispute involves issues arising in the banking industry, construction field, or international trade, the essential case management decisions remain consistent across disciplines. Similarly, whether the action is situated in a United State's Federal District Court or the Queen's Bench Division of England's High Court, the initial litigation management concern is whether the case will turn largely on legal issues or factual issues and then the size of anticipated document population which will result from discovery. How the legal issues can best be advanced and demonstrated, by the most accurate and cost-effective technology is of critical concern. The question is not only how quickly can a document be reproduced, stored and retrieved, but what methodology will most efficiently, expeditiously, and flexibly search through, locate, measure and prioritize the information and evidence located in that document.

This paper, then, looks at what automated litigation support system, and/or integration of systems, will best serve the needs of the particular case, remembering that electronic imaging is but one form of document reproduction, storage, and retrieval, and one element of a multi-faceted approach to find, review, value, and organize the information located in the documents.

Banking disputes may arise out of a complex of legal specialties such as the interpretation of contractual rights and obligations, securities investment and regulation, real estate mortgages, industry bankruptcies, fraudulent practices and any number of other complex financial transactions.

These conflicts nearly always involve multiple parties, extensive documentation, and numerous factual issues. Control over these factors may significantly impact the outcome of the case, making the need for fact management essential to any favorable outcome. In what documents are these facts found? Financial papers,

*V. Mital (ed.), Advanced Litigation Support & Document Imaging, 25–34.*
© 1995 *UNICOM Seminars. Printed in the Netherlands.*

security filings, investment agreements, business reports, meeting minutes, memos, letters?  Handwritten notes or marginalia comments by the major players?  The litigation support team's ability to select, evaluate and develop the crucial documentation of evidence is necessarily advanced by a detailed case plan.


## 5.1    FIVE BASIC STEPS TO DESIGNING THE CASE PLAN

### 5.1.1    Establish the Actual Goal(s)

Carefully identify and clarify the goals and develop the appropriate strategy.  Is the goal to defend the former officer(s) and/or director(s) reputation, no matter what, or to reach favorable terms for settlement?

### 3.1.2    Understand the Overview of the Case

Who are the parties involved, and what are their various relationships?  What are the plaintiff's legal issues, what are defendants' respective defenses, any counterclaims?  What potential damages or other redress is at stake?  Taking into consideration all anticipated parties, what is the extent of the document population, and how standardized are the documents' formats.

### 3.1.3    Establish "need to know" Facts

- Will a Statement of Claim be filed or will a solid understanding of your client's documents, together with other legal factors, be sufficient to achieve a favorable settlement without a legal proceeding?

- If an action is commenced, is trial likely?  Will the results of disclosure enable you to develop information critical for favorable settlement?

- What is the timeline?  How long do you have to review documents?  In England, discovery practices emphasize the exchange of documents, written interrogatories, and requests for admissions.  There is less reliance on oral testimony, which is prevalent in the United States.  This emphasis on making documents available for review and examination suggests the critical importance of being familiar with the precise contents of your documents and being able to effectively manage them.  Automated systems offer the prospect of acquiring in depth knowledge within rigid time restrictions.

- What are the discovery issues?  Is the case document intensive?  In what condition are the documents?  How much time will there be for analyzing opposing parties' documents?

- How many client documents need to be reviewed, organized, authenticated, protected as to privilege? What percentage of the documents are likely to contain information that is relevant or will lead to relevant sources of evidence. What percentage of opposing parties' documents will require careful reviewing, analyzing, refuting? How many and what kind of documents are likely to contain highly relevant information?

- What is the likelihood of sharing the costs of automation with other parties, subject to security considerations? For example, several defendants in a banking action may share common defenses, and strategies, in addition to defenses particular to each of them. Or, some of the data that is significantly helpful to one party may be significantly damaging to another. Can all parties be assured that their respective "work product" as to certain documents will remain secure from the viewing or use by other parties?

- How many lawyers will be involved, how large a support staff is available? Are they situated in one or many locations? What are their levels of computer literacy? If more than one firm is involved, what computer system compatibility issues are presented?

- Are there present, or future possible cases within the firm(s) with related factual or legal issues for this or other clients? Insurance companies in the United States often absorb the costs of defending former officers and directors of failed savings and loans. With a series of similar actions anticipated, these companies might be very interested in establishing a technology base for future application.

### 5.1.4  Develop a Case Outline

A thorough case outline focuses the team's attention on the critical issues involved in a complicated matter, so that the fact development and dispute resolution can be pursued in ways which will maximize impact and remain cost-efficient. Using a case outline allows the team to competently move through myriad documents and complicated events undistracted by irrelevant issues. The outline establishes a method for fact retrieval, and measuring progress. Such an outline can take advantage of the factual and legal issues developed in the course of the fact intensive pleading stages to confirm, eliminate and narrow the scope of the fact management involved.

### 5.1.5  Construct a Case Information Management Plan

By designing a plan to manage the information in the case, the team necessarily establishes the essential criteria upon which to base the merits of the case. Establishing this criteria early on, despite the inherent limitations, greatly enhances the ability of all involved to make informed, consistently focused decisions on discovery strategy and information management. The goal is to minimize the number of documents chosen for reproduction and prepare guidelines for how documents, once reproduced, will be indexed for retrieval purposes. What percentage of the documents are of value to reproduce at all? Of that amount, what reproduction methodology and indexing categories will be most appropriate?

The overall case plan, then, provides an integration of the client goals with a precise understanding of the legal and factual issues involved, the time restrictions imposed, and a clear focus for searching out the determinative facts as found in the documents of the case.

Additionally, the case plan process reveals the advantage, if any, of using a computerized management system. If the case looks like it will largely turn on facts; the facts are complex, to the extent that the documents will merit repeated review; and there is a large document population to organize, then computerization should be given serious consideration, if there is sufficient time. The advantage of the automated system is that you can control the documents more efficiently and cost effectively than through the more labor intensive and slower "by hand" methods. Secondly, a computerized system is inherently stable and is not detrimentally affected by any natural turnover of the litigation support team in the course of the case.

If a computerized management system is warranted, we must then turn our attention to what are the requisite components of that system. Can be rely on full text retrieval (FTR) or are the facts abstruse, requiring a skilled person to "read between the lines" and apply issue codes? Can we rely on objective coding criteria alone? Will some combination of objective and subjective coding, creating an abstract database, integrated with FTR capabilities prove to be the most cost-effective method? Once these decisions are reached we can then intelligently determine the value of EIM (electronic image management) versus the more conventional photocopy or microfilm options.

## 5.2    DOCUMENT REPRODUCTION

Imaging is the term used to describe the electronic process of scanning a document to capture and store its picture in digitalized form (usually) on an optical disk for future viewing on a high resolution monitor or reproducing through a laser printer. These images are often called, "bit-mapped", because it is the document's color gradations, represented as bits, that are being stored and recreated.[1]    A unique

---

[1]    Wm. E. Cwiklo, "Teaming Image Management Systems with Full-Text Databases", *Winning with Computers, Trial Practice in  the 21st century*, 1991, 254.

document control number, (what we typically refer to as a "litigation document control" or LDC number), is automatically assigned to each image as scanned. To later effectively retrieve that image, a document abstract should be created containing such objective information as the document type (letter, memo, or report), its date, title, author, etc.

Photocopying is of course the most familiar process. The essential equipment is an integral part of every firm regardless of size. The availability of seemingly limitless outside copying services serve to keep copying costs reasonable. Costs are dependent on volume and document preparation considerations. Cost advantages diminish as document populations grow. Why? Document control considerations require that each page reflect a unique identifying number (LDC) to maintain the integrity of the documents and facilitate accurate reference and retrieval. With photocopying, this numbering task must be manually performed, and labor is expensive. Also, it is important to maintain the integrity of the documents and their respective location within the filing system, another labor intensive endeavour. The danger of misfiling or otherwise compromising the system's integrity is high.[2]

Then too, customary case management practices dictate that the originals, or first generation copies of other parties' documents, remain secure, protected from destruction or loss, thus necessitating additional working copies, adding cost and increasing demand for extra storage capacity.

The advantages of microfilming are immediately obvious. A roll of microfilm, typically 3 × 3" contains approximately 2500 pages. Space efficiency is coupled with overall cost savings. Microfilm costs only pennies more per image with hardcopy reproduction (copyflo) additional. But, the camera can generate the important unique identifying number — the LDC number discussed above — for each page filmed, thereby combining tasks in a cost effective application. Once filmed, the integrity of the documents can only be compromised if the film is lost or damaged.

With either of the above methods, significant labor cost and time delays can be associated with retrieving needed copies. For example, you must manually search for, retrieve and reproduce all documents responsive to a database query. This task is generally performed by a case clerk but can ultimately eat up a lot of your budget. Either method requires each party to have its own set of documents or film or send an individual to the document depository to obtain needed copies.

Instead, visualize "looking" at the documents responsive to your database query on your computer monitor, and immediately printing a laser quality copy of any document of interest. Or responsive documents could be automatically printed

---

[2]   Studies show that there is a 3% misfile factor associated with paper-intensive operations.

in whatever sort order you choose. Most importantly, for shared database situations, electronic imaging allows for instantaneous communication of document based information and **multiple, simultaneous** viewing (or printing) of documents.

Attractive concept? Efficient? What is the real cost? One U.S. imaging service estimates the cost of scanning at $.15 to $.20 cents/page, with a penny per page for duplication. This cost is roughly twice to comparable microfilming services. Sequential LDC numbers are applied to each image and a rudimentary index to facilitate later searches for documents is included. But, the critical cost factor here depends on your pre-existing platform, i.e. what hardware/software combinations and capacities does your firm presently utilize. Currently, a single work-station, consisting of a microcomputer, high-resolution monitor, optical disk drive, scanner, printer and related software, can cost between 10,000 to 70,000 U.S. dollars depending on the configuration required.[3] With more than one firm involved, these baseline costs must be duplicated accordingly.

On the other hand, some imaging vendors, will incorporate the use of any necessary equipment into the pricing of their services, including essential workstations as part of its imaging package. Thus, for example, the document scanning for a case involving the imaging of a document population of 500,000 pages, and the need for three premium workstations, with in-house scanning capacity for additional documents, would be charged at a rate approximating $.25 cents per image[4].

Obviously, then, cost is an important factor in determining the appropriate reproduction methodology. Imaging may look more costly, but it contains real cost savings in terms of labor intensive "retrieval" of the actual documents. To know what is truly costly, and what is actually cost-effective, requires a thorough knowledge of the needs of the case.

## 5.2    DOCUMENT CODING

The decision of how best to reproduce documents in a particular case is but one aspect of the case management plan, providing as it does a particular way to **access** relevant documents. Of critical concern, however, is getting **information about the**

---

[3]    Digitized documents consume massive amounts of storage space, even when efficiently compressed. The optical disk component of your system is the most essential element. Further, it is hard to imagine that you would want other than a WORM (write once, read many) disk for evidentiary reasons.

[4]    Interview with Dave Tiller, Branch Manager, IPRO Inc., Phoenix, Arizona (February 14, 1992).

**document** — what does the text of the document say, who wrote it, who received it, who saw it, in what chain of command, who commented on it, etc. The functional ability to retrieve information is what most attracts lawyers to computerized litigation support systems.

Effective retrieval requires the creation of a database which captures pertinent information according to the dictates of the case plan. The appropriate litigation support staff then searches the database. If you have created an abstract database, the query will consist of a combination of objective attributes, such as "all letters authored by John Smith from May 1 through May 30, 1984". If you have coded subjective information, such as keywords or issues, evidentiary value of the document, etc. this criteria may be searched as well. The computer then provides you with a report of abstracts of all documents in the database satisfying your search parameters in a sort order of your choosing.

Typically, coding services are performed in-house by paralegal staff, or by an outside service. Because coding is generally the single most expensive line item on a litigation support budget, careful consideration should be given to the selection of what documents deserve this treatment.[5]

In a particular case, or for some portion of the relevant documents population, the advantages of FTR may be compelling. Instead of, or in addition to, a minimal abstract of each document, full text can be captured by utilizing optical character reader (OCR) technology. By digitally "reading" the characters of the text of the documents, OCR conversion automates, and theoretically streamlines, the information capture process to the obvious relief of already time-constrained lawyers. Secondly, OCR technology can be incorporated into the imaging process, allowing the lawyers to combine the information search and document access without additional database training.

OCR is a tremendously exciting technology where major advances may be expected within the next few years. However, at present, this application should be viewed with caution in the field of computerized litigation support. The caution here has to do with the accuracy of the text actually converted. OCR is best at recognizing computer generated laser quality print and standardized printed text, and requires more highly specialized, expensive equipment to reliably capture handwriting and graphics. The condition of the document itself is significant — is it first generation, wrinkled, old, faded? Proportional spacing can be a problem and columnar data will not be retrieved in its original format. Often special handling is necessary if the document contains a mixture of text, graphics and tables. All

---

[5]    In our experience, we have found that generally only 40-50% of all produced documents merit objective coding and only 40-60% of that population merits subjective or FTR treatment.

additional handling and specialized equipment add considerably to an already expensive proposition.

Cost is of course of paramount consideration and runs, generally $.50 to $1.50 per standardized page, assuming 1,000 character per page, with an accuracy rate of 95%. An accuracy rate of even 95% leaves questions about 1 of every 20 characters or perhaps 5 - 7 words a page. Accuracy drops dramatically when measuring the ability to capture handwritten comments, or other marginalia. Overseas high volume keying operations in places such as Taiwan, Korea and Jamaica, with verification controls, can be purchased at rates ranging from $1.30 to $2.50 per 1000 characters, but with accuracy rates at 99.5 percent[6].

If you contemplate utilizing OCR technology, then a pilot project consisting of an overall sampling of the documents under consideration is highly recommended. From the pilot you should be able to determine actual throughput figures, accuracy and costs, *inter alia*. A word of caution, however, the pilot is not always illustrative of how the overall project will progress. A number of considerations should go into your choice of an imaging and/or OCR vendor, one of which are the pilot project results.

OCR is, however, a truly amazing technology to watch. It will undoubtedly revolutionize the methods by which we presently search for, use and control information. It is fascinating to watch the technology as it grapples with its ability to "read" and constructs "dictionaries", "thesauri", and other techniques to combine the talents of artificial and creative intelligence to build in reliability and increase confidence.

With all of the variables being described in considering whether and how to best implement automated litigation support, it is essential that the decision-makers be in command not only of the complexity of the attendant legal issues but also be more than just conversant with developing computerized applications. With such dynamic, exponential development in the world of the computer, there are many obvious advantages to retaining the services of a consultant whose job it will be to meet your needs, by integrating the best choices from what is available. In so doing, the work of assessing a vendor's representations is done by a consultant whose only product is the needs of your case, satisfactorily met. That expert should be an integral part of the case management plan from the onset. In this way, the design choices for the document categories, the decisions as to using keywords and issues versus FTR can be more expertly combined with the often changing retrieval needs of the pre-trial team, and the litigation support process can be organized to accommodate the lawyers' approach.

---

[6]    Cwiklo, *Supra*, 257.

To determine whether an abstract database or full text retrieval will better serve the strategic goals of the legal action, it is important to consider the following factors:

- the extent of the document population, under your control, combined with the volume of documents likely to be produced and analyzed in the pre-trial stages

- what are the time demands, how quickly must the documents be reviewed and organized

- what documents are involved: mostly typed reports; banking industry forms, already extant in electronic format; handwritten memos, notes, or commentary

- what portion of the documents is comprised of photographs, budgets, designs, graphs, projections, etc.

- the nature and complexity of the factual disputes

- the accuracy of the OCR technology available

- the quality of the coding and indexing personnel and procedures

- the skills and experience of management personnel: their actual time commitment to the project

- the editing and quality control procedures already in place

- what is the present capacity and flexibility of the firm's computer equipment, hardware and software

- what is the real cost savings in terms of time and labor as compared with the expense of implementing the respective methodologies.

It is factors such as these, considered together with the experience of the consultant, or resident expert, which will determine what systems, or integrated combination, will serve your client's best interests.

Considerations such as these must be measured against the particulars of a given case to determine what management system or integration of applications will be selected. Making these decisions with the help of a consultant will allow you to select abstract coding or FTR as appropriate to the actual documents, and the most cost-effective retrieval method really needed. Most importantly, working jointly with a consultant helps you merge the expertise of the two fields of law and computer science to provide the most efficient and effective service to your present client while designing the firm's future stability and expansion.

But, whether you are considering moving on from current manual information management methods to computerized systems or enhancing your current technology to embrace EIM and/or OCR, I suggest that you not overlook the question of whether or not your users, be they barristers, solicitors or support staff are willing to embrace the new technology and dramatically change the way they currently do business. If they are not, the technology may be for naught.

# 6. Applying Imaging: A Survey of the U.S. Law Office Scene

*Joseph M. Howie, Jr.*
*Docucon Inc*

## 6.1    OVERVIEW OF THE U.S. LEGAL MARKET

The U.S. legal system is still in the throes of a socio-economic revolution much like that facing the European Common Market.  In the United States, the practice of law was for centuries a "learned profession" in which it was deemed ungentlemanly or unprofessional to overtly compete with your brethren.  Bar Associations set fee schedules and had ethical prohibitions on advertising.  However, in *Goldfarb vs. Virginia State Bar*, 421 U.S. 773 (1975), the United States Supreme Court held that the Sherman Antitrust Act applied to price fixing by bar associations, and in *Bates vs. State Bar of Arizona*, 433 U.S. 350 (1977) the Supreme Court said that the First Amendment to the United States Constitution included a right to free speech that protected some forms of lawyer advertising

Since then, the legal marketplace has indeed gotten more competitive.  While the top commercial firms do not advertise on television or on billboards like the personal injury or workers compensation firms, they do have firm brochures and they have marketing committees that talk about things like business generation and cross-selling of existing clients.  Technology and information management are marketing weapons being employed by the top firms in their battle for clients.  Firms today want to be more efficient and they want to be perceived as being more efficient.

Imaging is a frequent topic of discussion at legal technology conferences in the United States and in the legal technology literature.  Thus far, though, there has been far more interest than implementation.  Where imaging has been implemented in law offices it has tended to be in larger, more sophisticated firms which first implement imaging in the area of litigation support.  The reasons for implementing litigation support first relate to the nature of the discovery problems in the U.S., the ability of firms to charge back many of the image processing costs to clients, and the amenability of litigation support to being initially implemented on a less expensive stand-alone (i.e. non-networked) basis.  This survey of imaging in U.S. law offices

*V. Mital (ed.), Advanced Litigation Support & Document Imaging, 35–52.*
© 1995 *UNICOM Seminars. Printed in the Netherlands.*

therefore requires an examination of "litigation support". Understanding how imaging works in litigation support will provide insight into other potential legal imaging applications.

## 6.2    "LITIGATION SUPPORT"

To understand what "litigation support" means and why it is the most frequently implemented imaging application, we will examine certain aspects of the U.S. legal system and some of the associated economics.

**Discovery in the U.S. Legal System.** In the United States there are two parallel court systems, the Federal courts and the courts of the 50 states. Both court systems hear two kinds of cases, criminal and civil. In civil litigation parties can conduct pre-trial "discovery" to find out what the other parties or even non-litigants know about the facts of the case. For our purposes the two most important forms of discovery are "document requests" and "depositions".

**Document Requests.** By using document requests, a party can review and copy documents in the possession or control of another party. The standard used by courts to determine whether a party needs to comply with document requests is whether or not the requests are designed to obtain documents that are either relevant or likely to lead to relevant information. An example of a document request might be as follows:

> "All documents that refer or relate to the production of
> crude oil in California, or Texas in the 1970's."

As can be imagined, the volume of documents that can be requested can be staggering, sometimes calling for practically every document generated or received by large companies over long periods of time.

**Sanctions.** If a party knowingly fails to produce documents that are responsive to a document request without objecting to their production, that party and possibly that party's lawyers can face severe sanctions ranging from monetary fines, imposition of attorney fees, being precluded from asserting certain defenses, or possibly being precluded from being able to contest liability. The potential for sanctions greatly increases the financial stakes associated with complying fully with document requests.

**Multiple Jurisdiction Productions.** In the United States the discovery problem is exacerbated by the fact that a company often faces similar litigation in several jurisdictions, e.g. a manufacturer may face product liability claims in several states arising out of alleged defects in a particular model product. Plaintiffs' lawyers collaborate with each other to compare the documents they each receive from the same defendant in the hopes of uncovering discrepancies and hence potential sanction situations.

**Litigation Support.** The term "litigation support" or as it is sometimes called, "automated litigation support", describes the process by which computer systems are used to create indexes or databases of the documents involved in cases and to then search them to locate documents meeting certain search criteria. The most common use of a litigation support system occurs when notice is given that a witness will have his or her "deposition" taken. (A deposition involves having the witness sworn to tell the truth and then being asked questions by the attorneys for the parties with the questions and answers recorded stenographically by a court reporter. Depositions are typically conducted in a conference room at a lawyer's office without the judge or a magistrate present. During the deposition the witness is questioned about documents that are made exhibits to the deposition. After the deposition, the transcript of the proceedings and copies of the exhibits are produced for dissemination to the parties.)

**Witness Briefing Books.** Most lawyers who take or defend a deposition want to first review the documents that might be shown the witness, such as ones that were authored by, addressed to, or which mention the witness. It is very expensive to have paralegals and document clerks manually review all the documents to select the ones containing the name for each witness.

**Semi-Automated Litigation Support.** Most large firms nowadays have what is really "semi-automated" litigation support, i.e. they use some form of database software to store and access indexes to the documents or copies of the documents, but when it comes time to review or copy the documents, that is done manually working from paper copies or microfilm or fiche. In semi-automated systems, the documents are numbered sequentially and kept in document number order. When searches are conducted, the systems produce listings that include the numbers of documents to be pulled for copying. The documents are pulled from the shelves or filing cabinets, copied, and refiled. The copies are usually three-hole punched and put in three-ring binders for ready reference, arranged in chronological order.

**Image-Based (Fully Automated) Litigation Support.** In an image-based litigation support system, the documents to be processed for discovery purposes are scanned using image scanners instead of being copied in a paper format. The scanning is typically done at a resolution of 300 dots per inch. That is about the resolution to which lawyers are accustomed from their office paper copiers. Images scanned at 100 or 200 dpi will be noticeably coarser, somewhat like fax copies.

**Image Storage.** Each page that is scanned is stored in a compressed form in a separate computer file. Each page-image file is about 50 kilobytes (KB), i.e. about 20 pages per megabyte (MB). The large size of the image files mandates, as a practical matter, the use of optical disk storage for the files. Unlike hard disks or floppy disks, optical disk drives use tiny laser beams to write data onto the optical disks. The laser beam etches a layer of plastic in the disk, causing the spot that is etched to reflect light differently from spots that have not been etched.

**Indexes to Document Images.** Without an index or database to permit rapid access to the appropriate images, imaging is nothing more than electronic microfilm. In other words the image itself is not searchable. Meaningful access to the images requires the creation of a database that will permit the identification of the relevant pages/images. The database itself is typically kept on a hard disk. The database can be created in a manual coding process in the traditional, structured form, e.g. Date, Source, Document Type, Title, Author, Addressee, etc.

**Optical Character Recognition.** As an alternative to a manually-created index, image files can be processed through optical character recognition (OCR) systems to produce searchable text files, depending on the quality of the original documents. Many indexing variations are possible, including abstracting some documents, and OCR'ing the balance, or abstracting all of them with additional OCR treatment of a subset of the documents. Part of the challenge to today's litigators is to understand the advantages, limitations, costs, and reasonable expectations for the various types of indexing that are available.

## 6.3    RELATIVE ADVANTAGES OF IMAGING IN LITIGATION SUPPORT

Imaging has several generic advantages that are present in the litigation context:

**Compactness.** One 5.25" optical disk can hold from 10-12,000 pages. One 12" optical disk can hold from 100-120,000 pages. By comparison, one small document box will hold about 2,500 pages and one large document box will hold about 5,000 pages. A four drawer filing cabinet will hold approximately 10,000 pages. When document collections are in the tens or hundreds of thousands of pages, the savings in space can be significant.

**Archival Integrity.** According to studies by AIIM (the Association for Image and Information Management), when documents are pulled from large document populations three percent of them are usually lost or misfiled. In an imaging system, retrieval needs are satisfied by searching the database and viewing or printing image files, and the original documents do not need to be disturbed, thereby avoiding the impact of this "three percent rule".

**IPG.** In a manual or semi-automated system, it may take hours or days from the time that an information request is made by a lawyer and the time that paper copies of the responsive documents are available for the lawyer's review. By that time there may no longer be the opportunity to use the information, e.g. the witness may no longer be available, or the case may be over. With an imaging system, any document out of millions of pages can be viewed and printed within literally seconds. This "Immediate Positive Gratification" not only makes the underlying information more available, its ease of use encourages using the system, improving the cost-to-benefit ratio of the system. After all, the worst cost-to-benefit ratio occurs in a system that is so hard to use that it never gets used and therefore never provides a benefit.

Additionally, there are some benefits to imaging that are somewhat unique in a litigation support context:

**Decreased Cost of Briefing Books.** With a manual or semi-automated system, it may take two paralegals literally days to create a briefing book for just one witness. With an imaging system however, briefing books can be created on even a low-end system at the rate of 300 pages per hour, saving literally thousands of dollars of legal fees for each witness.

**Sorting of Blow-backs (Paper Copies).** Many times it is helpful to attorneys to have sets of documents printed out or blown-back in a particular order other than the order in which they were initially scanned. For example, for witness briefing books, most lawyers want the documents presented in chronological order to help them get a sense of how the facts developed. With an image-enabled database, the documents can all be sorted prior to printing so that a paralegal or document clerk does not have to sort the paper copies. For other purposes it may be helpful to review documents that have been grouped or sorted by document type or by the source of the document.

**Decreased Copying Costs.** In litigation, there may be a need to make several complete sets of copies of the documents, e.g. one set for corporate counsel, and one set for each of the regional counsel handling specific cases. In the U.S. each set may cost $.150 per page per copy. With imaging systems, the disks can just be duplicated, at a cost of from $.005 to $.015 per page per set. The savings for just making copies can pay for the scanning, indexing, and duplication of the disks.

**Decreased Shipping Costs.** In a litigation setting, each copy of the set of documents may have to be shipped to a different location, typically using some sort of expedited shipping service. With imaging, the optical disks can be sent for a tiny fraction of what the paper set would have required.

**Document Numbering.** It is common to number each page of the documents that are produced to the other side, typically through the application of labels that have been numbered with computer-generated numbers. The cost of creating and applying the label will be $.05 or more per page. In an imaging system, the laser printer can print the numbers on the pages at the time the documents are printed out, thereby avoiding that cost all together.

**Reproducing the Whole Page.** When documents are copied using normal office paper copiers, there is usually a margin around each edge of the paper that does not get copied. To see how large that margin is on your copier, take a piece of paper that is completely dark and copy it. You may find that there is about a quarter to a third of an inch all the way around the paper that does not reproduce. In most settings that may be acceptable, but it may not in litigation support where evidentiary documents are being processed. In an imaging system, the documents can be slightly reduced when they are reproduced so that the entire document image can be seen on the page, possibly even reduced so that when the paper is three-hole punched, none of the image is lost.

Two of the primary factors in law firms selecting litigation support as the first imaging application are: the facts that (1) the client may in effect pay for a firm's first test of an imaging system, and (2) imaging can be implemented on a less expensive stand-alone basis.

**Hardware Requirements for Litigation Support.** In litigation support it is possible to have a very effective imaging implementation using a stand-alone PC system, i.e. one that is not on a network. In a stand-alone environment, paralegals can use a image-enabled PC to search a database, find the relevant documents and then print them out for the lawyers. From the lawyer's perspective the only thing that changes with a stand-alone system is that the paper comes to his or her desk more quickly. Ignoring for a moment the cost of a scanner, a firm or corporation can have a working imaging system by adding less than $8,000 worth of hardware to an existing 80386 PC. With Windows the cost will be only half of that.

In a stand-alone system the PC operator is prompted by the computer screen when the system needs to display an image that is on another optical disk. Because the operator is sitting in front of or next to the optical disk drive, that can be done without much delay or inconvenience. When images have to be distributed over a network, the operators will not have direct physical access to the optical disk drive, and the use of a jukebox is required. A "jukebox" is an electromechanical device that will store multiple optical disks and then mount and dismount the necessary disk as prompted by the computer. Jukeboxes are typically the most expensive of the various imaging components that are used.

**Client Chargebacks.** A U.S. law firm typically charges document processing costs back to clients, and that practice continues when the firm switches to imaging. Most firms will bill clients for scanning and indexing documents. Firms will vary in how they attempt to recoup their hardware costs, e.g. scanners and jukeboxes. Some firms feel that they should treat those type of capital expenditures as part of their overhead and not bill separately for them. Others implement imaging at the behest of a single client for a specific matter and they may bill all the costs back to the client.

## 6.4   BATCH-ORIENTED LITIGATION SUPPORT

There are two kinds of imaging systems, batch-oriented, and process flow. In a batch-oriented system, the bulk of the documents being managed are scanned and indexed at the outset and for the most part imaging is used to manage the documents that have been archived on optical disk. In a process flow application, imaging is used to manage an ongoing, relatively steady stream of documents for the purpose of processing them or making decisions about them. Imaging is used to facilitate the rapid electronic dissemination of the documents to the various decision makers who need to review them.

Litigation support is for the most part a batch-oriented application when used for large document cases. Examples of process flow applications include insurance company claims management and mail distribution systems. Process flow almost by definition will involve multi-user networked system, meaning that jukeboxes will be used and there will be multiple image-enabled workstations involved.

From the law firm's perspective, the fact that litigation support can be a batch-oriented imaging application can present certain advantages when first trying imaging. Because there will not be an ongoing need for the hardware and personnel needed for scanning and indexing, those services can be "outsourced" or purchased on a per page basis without the firm making a commitment to hire permanent staff or risk major investments in technology.

Even when the firm decides to have an ongoing image scanning and indexing capability, it may well decide to outsource for major peaks in workload caused by large document production. Additionally, the fact that many documents may be produced in remote locations may suggest the use of vendors in those remote locations.

## 6.5   COLLABORATIVE WORKING, CASE ANALYSIS
##         AND ANNOTATION OF EVIDENCE

In large document cases, it is not possible for a lawyer to look at every piece of paper produced. Large document cases are team efforts, and where image-based litigation support is available in a networked environment it represents an unparalleled opportunity for collaborative working, case analysis and annotation of evidence.

In the above section we discussed how it was very feasible and beneficial to implement imaging on a stand-alone basis in a litigation support setting. Documents are available faster, they can be sorted prior to printing, and the problems of missing documents are eliminated. The lawyers and paralegals who are the end users of the paper are provided with the paper in a quicker, more reliable fashion. However, having all members of the team on-line with networked imaging automates the way they work, not just the way that the support staff works.

Networked imaging lets the firm use the litigation support database as an institutional memory bank of the team's insight and analysis of the facts and documents in the case. There are two specific techniques that can be used in an on-line imaging system to build the collaborative work product of the team in a structured, predefined manner - issue coding or relevance coding of the documents. While either concept could be used without imaging, having distributed imaging enables them to be implemented with far less complicated and far less burdensome work procedures because the lawyers and paralegals have both the images and the relevant database records displayed at the same time on their terminals.

**Issue Coding.** Issue coding involves identifying of certain documents that pertain to particular issues in a case. Rather than relying on people to use the same vocabulary in narrative comments, an issue coding procedure requires people to use a controlled vocabulary that enables more precise recall of desired documents during retrieval. Lawyers have a well developed ability to express the same or very similar thoughts in numerous ways, and while that may be considered a fine art when drafting contracts or pleadings, it makes for very imprecise retrieval. For example, when databse users search a database for documents where the issue "Statute of Limitations, Tolling" occurred, they will not find an entry where someone made up his or her own abbreviation, e.g. "SOL-T" or "Tolling" or "Fraudulent Concealment" or "FC" or "Fraud".

With imaging systems, users have the unique opportunity to be viewing the actual image of a document at the same time that they have immediate access to the computer record that indexes that document. With a standard vocabulary and a structure to the issues in the case, the lawyers and paralegals in the case can immediately identify the significance of the document so that it does not become lost amidst the thousands of other records.

**Relevance Codes.** Relevance coding has the same general purpose as issue coding —the identification of the significant documents in a case. However, rather than attempt to construct a detailed issue code table and/or controlled vocabulary for coding of documents, relevance coding is a simpler system of noting the relative significance of documents. A simple example would be to use a scale of one to five where all documents are presumed to have a significance of "1" to start out, and are assigned higher significance levels as the case progresses. For example, "5" may indicate a document that could be dispositive of the entire case; "4" one that may dispose of a major issue; "3" a document that should at least be a trial exhibit; and "2" a document that may be a deposition exhibit. People who favor relevance coding over issue coding feel that it is impractical to construct precise issue code tables at the outset of the case, and the really significant documents end up having multiple issue codes anyway.

**Validation of Subjective Coding - Avoiding the Garbage In, Gospel Out Syndrome.** Whether a lawyer decides to implement issue coding or relevance coding, that lawyer needs to validate the manner in which the coding is actually being performed. To do that, provide the same set of 15 or 20 documents to the people who are doing the coding and ask them to code the documents. Tabulate the results to see if the coders are all assigning about the same number of codes to the documents (i.e. the "intensity" of the coding), and whether they are assigning the same codes (i.e. the "congruence" of the coding). Don't invest hundreds of hours of billable time before testing the consistency of the coding. If it is so inconsistent that you can't rely on it and won't use it, you should learn early enough to avoid it altogether. The worst case of course would be to have subjective coding performed and then rely on the results as if they were absolutely correct when in fact they were very inconsistent.

**Hardware Considerations.** For distributed or departmental imaging to be effective, the minimal requirement is a jukebox to operate as an image server on the network. If the network is operating under Novell or Banyan and the workstations are PC's capable of operating Windows (i.e. they are 80386's with at least 4 MB of RAM), the basic hardware is already in place, unless the firm chooses to go with proprietary imaging hardware, but that seems to make about as much sense as purchasing proprietary word processing equipment instead of PC's.

## 6.6    SUPERVISING ASSOCIATES AND "PARALEGALS"

**Lawyers As Coaches.** As has been perhaps implicit in much of the above discussion, a lawyer who is responsible for a large document case must do more than "lawyer". To obtain good results in such a case the lawyer must function as much as a manager and a coach as a lawyer. Increasingly law firms are employing or retaining the services of people who have specialized knowledge and skill. While it is not the lawyer's job to know everything, it is his or her job to know what he or she doesn't know and to know where to go to find people with the requisite knowledge. Clear delegation to staff members and the accountability of the staff to defined standards contribute as much to the success of the litigation effort as pure legal knowledge or technical expertise.

**Handling of Large Cases.** It may be helpful to briefly review how a large document case might be staffed in the U.S. Typically when a large corporation is sued in the United States, the first lawyer to learn of it is someone in the in-house legal department. The corporation will usually retain a law firm or outside counsel in the same city or region where the law suit was filed, and a partner of the firm will assume overall responsibility for the case. Associates will generally be responsible for initially drafting pleadings and taking the depositions of the lesser witnesses. The partner will take a hand in completing the drafting of more significant pleadings, arguing motions to the court, taking the depositions of significant witnesses, and in the cases that actually go to trial, in being lead counsel in trying the case.

**Paralegals.** Paralegals are invariably used for much of the document production work, i.e. locating and reviewing the documents to be produced. Paralegals generally have at least a four year college degree and/or specialized paralegal training that will include overview courses on civil procedure, contracts, torts and legal research. Their billing rates are generally lower than those of associates, and while their work must be supervised by an attorney in some fashion, their use can result in lower legal fees for the client, and increased job satisfaction for the young associates who would otherwise be tasked with doing fairly mundane and tedious work. It also permits the lawyers to handle more cases. As a result of their hands-on review of the documents in the case, paralegals may end up knowing the documents and the facts of a case at least as well as the partner who is responsible for the case.

When an imaging project is undertaken it is usually the paralegal who gives day to day direction as to scanning and indexing of the documents. The senior paralegal will be the one who has contact with the corporate client and arranges the details concerning on-site scanning at the client's site. Associates or partners will be involved in arranging the details of document scanning that takes place at locations that are operated by adversarial parties.

**Project Design Manual.** As can be seen, an image-based litigation support project typically involves the downward delegation of responsibility from the partner handling the case to the associates who will do much of the briefing and deposing, to the paralegals who do much of the document reviewing and production work, and to the litigation support section or vendor who does the scanning and the database creation work. (Note that it is not appropriate for persons with the education, experience and capabilities of a litigation paralegal to be actually scanning or doing data entry work.) In image-based litigation support those different levels and types of people communicate through the creation and dissemination of a Project Design Manual that sets forth the volume of documents to be processed, the priorities for different categories of documents, the coding conventions to be used, the type of system to be delivered, who is to be provided access to the database and images, and so forth. When the project is completed the Project Design Manual will also contain explanations about any deviations or enhancements that were made to the initial Project Design Manual.

## 6.7     COST CONTROL BY CORPORATE LEGAL STAFF

**Former Incentive to be Inefficient.** Law firms have had a perverse economic incentive to be inefficient: each hour spent being inefficient was a billable hour. No more. Corporate counsel are increasingly aware of what document management costs are, and what the technological options are. Law firms which take a long term, strategic view of their planning recognize that it is now in their own economic best interest to be efficient in the processing of documents. The reason is obvious: law firms make more money selling legal services than they do selling document copying or indexing services. Being perceived as being deliberately or inadvertently inefficient in document handling ends up costing law firms legal business.

**Document Management Cost Control.** In-house corporate staffs in the U.S. are under increasing pressure from corporate CEO's and Boards of Directors to lower or control legal fees. Corporate counsel are expecting detailed bills and are questioning each line of detail. One of the areas getting increasing focus is the document management part of litigation. U.S. law firms formerly billed $.20 or $.25 per page for the copying of litigation documents. In fact, the cost capture systems on office paper copiers used to pay for entire legal automation systems. Corporate clients are now coming to expect law firms to use more efficient, less costly outside services to provide copying.

**Bringing Litigation Support In-house.** This corporate focus on controlling document management costs is extending to the cost of creating the initial database that indexes the discoverery documents. Some corporations have brought this function in-house so that they do not get billed normal paralegal hourly rates for document indexing by their law firms. Carl Liggio, the general counsel for Ernst & Young has indicated that he planned on supplying certain outside counsel with an image-enabled computer system, document images on optical disk, and a completed database to use in accessing documents, because he felt that Ernst & Young would save legal fees in the long run if their outside counsel were automated.

**Contracting Separately for Litigation Support.** Other corporations are contracting for document scanning and indexing from companies that specialize in such document conversion work. This not only diverts revenue from the law firm, it gives control over the information in the case to the corporation instead of the law firm. That makes it easier for the corporation to replace outside counsel or to use its in-house staff to coordinate with regional counsels who are handling similar litigation in different regions.

**Cost Analysis of Document Processing.** The following worksheet suggests the kind of analysis a corporate counsel may perform in deciding how to handle the documents associated with a 200,000 page case. It demonstrates that the corporation can receive lower cost, better quality document management with imaging.

## COST WORKSHEET:
## IMAGING VS. LAW FIRM-CREATED INDEX WITH PAPER COPIES

Note: This example is based on numbers that would be encountered in a 200,000 page case where two sets of documents will be made (perhaps two law firms, or one in-house counsel and an outside counsel). Actual numbers may and probably will vary somewhat. ($US).

| Numbering | Manual | Image |
|---|---|---|
| Cost to affix labels or stamp numbers<br>Imaging - show on printouts | $.05/page | $.00/pg |
| Net Savings - Numbering | | $10,000 |

| Copying | Manual | Image |
|---|---|---|
| Cost to copy first set paper<br>Cost to copy second set paper | $.15/page<br>$.10/page | |
| Cost to scan first image<br>Cost to make second set of opticals | | $.20/page<br>$.01/page |
| Total cost, two sets of copies<br>Net Savings, Copying | $50,000 | $42,000<br>$ 8,000 |

## Cost Worksheet (continued)  Manual  Image

### Storing

| | Manual | Image |
|---|---|---|
| Number of four drawer file cabinets needed for one set | | 20 |
| Cost per file cabinet | | $125 |
| Cost for file cabinets | | $2,500 |
| Pages per manilla folder | | 100 |
| Cost per manilla folder | | $.07 |
| Number of manilla folders | | 2,000 |
| Cost of manilla folders | | $140 |
| Square feet per cabinet | | 7 |
| Total square feet | | 140 |
| Cost per month/square foot | | $.80/month |
| Estimated life of case | | 24 months |
| Cost of space for 24 months | | $2,688 |
| Total cost of storing one set | | $5,328 |
| | | |
| Cost of storing two sets paper | $10,656 | |
| | | |
| Imaging Upgrades to Two PC's | · | $8,000 |
| | | |
| **Net Savings, Storing** | | $2,656 |

### Indexing

| | Manual | Image |
|---|---|---|
| Temp Agency Personnel Hours To Index | | |
| 200,000 pages @ 25 pgs/hour | 8,000 hours | |
| Cost per hour | $30 | |
| Total Cost to Index w/Temps | $240,000 | |
| | | |
| Cost to have Vendor Index | | $.70/pg |
| Total Cost to have Vendor Index | | $140,000 |
| | | |
| **Net Savings, Indexing** | | $100,000 |

### Retrieval

| | Manual | Image |
|---|---|---|
| Total Number of Witnesses to be Deposed (or other retrieval projects) | 20 | |
| Average Number of Pages per Witness for Witness Kit | 900 pgs | |
| Number of Pages to be Printed Out: | 18,000 | |

| Cost Worksheet (continued) | Manual | Image |
|---|---|---|

### Retrieval (continued)

| | Manual | Image |
|---|---|---|
| Time to Manually Copy, Refile and Distribute Witness Briefing Books | 30 hrs/witness | |
| Total Hours Manually | 600 hrs | |
| Cost per hour | $50/hr | |
| Total Assembly Labor Costs, Manual | $30,000 | |
| Copying Costs/Page | $.20 | |
| Total Copying Fees | $3,600 | |
| | | |
| Time to Assemble Witness Kits Using Imaging hrs/book | | 3 hrs |
| Total time, w/Imaging | | 60 hrs |
| Cost per hour | | $60 |
| Total Assembly Labor Costs, Imaging | | $1,800 |
| Copying Cost/Page | | $.10 |
| Total Copying Costs | | $1,800 |
| | | |
| **Net Savings, Retrieval, 20 times** | | **$30,000** |

TOTAL SAVINGS USING IMAGE PROCESSING
COMPARED TO MANUAL SYSTEM........................................................$150,656

No law firm today would bill clients for time spent by secretaries using manual typewriters to revise drafts of briefs. Imaging helps firms automate their discovery document management process like they have automated their word processing system.

Other than bringing litigation support in-house or contracting for it separately from legal services, there are two cost saving techniques using imaging that are receiving more attention.

**Sharing Databases and Images With Co-Parties.** In the U.S. it has long been customary for joint defendants to create a litigation support database for a case, sharing the costs of creating it and providing access or copies of the database to those who participated in its creation. Because images are "transportable" to different systems by converting one image format to another, the cost sharing precedent can be extended to enable co-parties to each obtain the data and the images in the format required by their own system.

**Sharing Databases and Images With Adverse Parties.** I suspect it will become more common for adverse parties to share in the cost of scanning and having bibliographic coding done by a third party. It does not make economic sense for both parties to physically scan the same documents and then to have both of them create databases that include entries for objective bibliographic entries like Date, Source, Type, Title, Author, Addressee. Note that in the U.S. the sharing of the objective bibliographic data would not affect the attorney work product protection against discovery that would still attach to the subjective or other coding that was done to the documents and the database while being used by the lawyers and paralegals.

**Avoiding Sanctions With Pre-Production Databases.** More corporations are attempting to avoid sanctions for inconsistent or incomplete document productions by creating and using an image-enabled database for the documents that pertain to a particular class of case, e.g. products liability or toxic tort. In other words, rather than relying on several regional counsel to review documents and make their own subjective decisions about what to produce, and then to use different systems of tracking what was produced, corporations are creating image-enabled databases for responding to document requests. This lets them keep a permanent record of what was produced, when and to whom. It also lets them print unique identifying numbers on each page-image that is produced to opposing parties.

## 6.8    CLIENT FILE MANAGEMENT

One logical use for imaging systems is to manage case files and the documents relating to them using images rather than paper documents. In such a system, the incoming correspondence and pleadings in a case would be scanned, indexed at some level, and then routed electronically to the responsible lawyer or paralegal. The originals could be placed in a master case file for archiving purposes, or, to save the cost of maintaining dual paper and electronic files, all incoming paper could just be scanned and placed sequentially in storage files, with primary reliance being on the electronic files. The idea would be that the paper records could be available in the unlikely event they were ever called for, but they would not have to be maintained on the premises of the law firm. Client file management using imaging system will be a logical outgrowth of having imaging hardware, software and procedures in place, and having the trained staff members who know how to use them. Note that client file management differs from litigation support in several important aspects:

1.    Client file management is a "process flow" application, and is not batch oriented. In other words, scanning needs to be done on an ongoing, real-time basis. In litigation support it may be adequate to have the documents scanned and have an index for accessing the images several weeks or possibly even months later. Therefore, as a practical matter the firm implementing client file management will

have to have scanning hardware and some type of indexing staff in-house.

2. Client file management is an inherently multi-user application. Unlike litigation support, firms will not be able to use stand-alone PC's effectively. This means that a jukebox is a necessity.

3. Client file management will require a deeper level of commitment to imaging, since nearly everyone in the firm will or should make the changeover to imaging as the preferred method of dealing with document-based information.

Insurance companies are implementing image-based claims management systems that are similar in function to what could be achieved with case management systems. USAA in San Antonio, Texas, for example was one of the first companies to implement an IBM imaging solution for its claims processing.

## 6.9    WHAT KIND OF FIRMS USE IMAGING

With litigation support being the most prevalent imaging application, it should not be surprising that most of the firms that have implemented imaging solutions are firms that are heavily litigation oriented, with an emphasis on the commercial types of disputes that generate voluminous document productions as part of the discovery process. They are firms that have, in some form, strategic business plans that rely on technology and information management to differentiate themselves in an increasingly competitive marketplace.

Imaging is still considered a "leading edge" type of application in the US. In other words, imaging is still "news" and there are many lawyers who have still not seen an imaging system in operation. On the whole though, lawyers and paralegals who are charged with the responsibility for managing documents relating to litigation are familiar with the concepts of imaging, have seen several products demonstrated and are actively considering their initial projects or systems.

## 6.10    ELECTRONIC DISCOVERY

Lawyers charged with conducting discovery against other parties need at least a minimal level of familiarity with how those records are kept. As more and more businesses move in the direction of keeping documents in electronic image format, lawyers will be forced to learn at least some of the concepts and terminology involved. Otherwise they will be missing a significant opportunity to learn the facts in the case and to acquire the information that is the most understandable with the least data processing costs. In other words, even if you think that you don't want to implement imaging at this point in time for your firm, you need to understand how your clients' or your adversaries' imaging systems work.

The need to learn about electronic media discovery for document images is essentially the same need that litigators have to be alert to opportunities where they can discover the electronic data of their adversaries, either because the electronic data is more useful than paper documents (as in word processing files that have time and date references that are not apparent on the paper copy), or it is the only evidence of a communication (as in many e-mail systems, or where an adversary claims that all paper records were destroyed), or where it would simply be easier to search and store than would paper records.

## 6.11   PROBLEMS WITH IMAGING

**Hardware Compatibility.**   Until recently, implementing imaging meant buying proprietary hardware or computer systems that required operating systems that would not otherwise have been used in the firm, e.g. Unix. The growing acceptance of Windows as a standard for imaging systems appears to be solving many of those problems.   There remains the problem that imaging components have not reached the maturity of stereo systems where one can buy a receiver and some speakers and a CD player and not be too concerned about having each component able to work with the others.   New purchasers need to be certain that certain key hardware components will work with each other.   Those components include:

- Optical disk drive
- Controller for optical disk drive or jukebox
- Interface card to connect PC to optical disk controller or jukebox controller
- Image expansion card for displaying images (may not be necessary if software expansion is used)
- Video card for displaying images
- Image expansion card for sending expanded images to printer
- Laser or ion deposition printer
- Computer memory for operating the imaging software.

In other words, just because you have a PC-based imaging system does not mean you can just load another imaging application and have it work properly.

**Printing.**   Printing deserves special mention.   While the promise of imaging has been the "paperless" office, in a litigation support setting, an imaging work station may actually end up being used as a high volume print station.   You need to take a careful look at your document reproduction needs and make sure your printer has the rated capacity to handle those needs in your time frame.

**Lack of Functionality.**   There are also some systems being touted as image-based litigation support systems that simply lack the needed robustness.  One package claimed to offer imaging but the images were uncompressed, 70 dpi images, and it did not support optical disk storage of images.  That meant that the images were of a completely unsatisfactory quality and valuable hard disk space was required to store them.  It was in reality more of a toy than a tool.  Other supposedly image-enabled software does not have some essential features like Boolean searching or multiple entry fields for data elements like Authors or Addressees.

**Improper Image File Structure.**   As another example of potential problems, some systems rely on storing the image files in a DOS file format, and while that may work fine in a small volume environment, when the firm goes to larger volumes and wants to use a jukebox, the DOS file conventions may not be adequate.   One "consultant" suggests a litigation support system that stores image files in a DOS file structure, but for performance reasons, prefers to have less than 300 pages in a subdirectory.  That means that on a one million page case, there will be over 3,300 subdirectories!

**Over-Reliance on OCR as the Exclusive Indexing Tool.**   Some users have also experienced unhappiness with the technique of relying on OCR'd full text as the exclusive means of searching a litigation support database.  They have discovered that the OCR process may not successfully convert the documents or the text in which they are the most interested, and that the cost of editing the text to bring it to satisfactory standards is prohibitive.   Other problems they have encountered relate to not being able to sort the output, and not being able to have formatted report capabilities.  Part of the problem lies in having unrealistic expectations for OCR'd full text based on experience with on-line systems like Lexis or Westlaw where the data was double-keyed and verified by having been used by hundreds of thousands of lawyers.  Expectations are also unrealistically inflated by experience with court reporter-generated text files that are 100% accurate.  OCR'ing is simply not that accurate.

# 7.  The Document Worker and Productivity

*Graham Pearson*
*Rank Xerox (UK) Ltd*


## 7.1    INTRODUCTION

Hello, and welcome to the Document Imaging in the Law Office Conference.

Rank Xerox as "The Document Company" has to express an interest in this subject as we are already a major supplier of document technologies to many firms engaged in the legal profession.

The main purpose of my presentation is to attempt to put into context the current state of technology, and some of the technical changes we see happening over the next few years as well as to develop some of the implications that these changes may have on the way in which organisations apply technology to gain the competitive edge that will underwrite their success in the future.

In attempting to address a group of professionals from a discipline other than my own it is important to first define some terms, so that we can establish a common language to assist understanding.

In the course of this conference we will be using the term document. In Rank Xerox we devote a great deal of time to the understanding and management of documents. At the outset of any discussion on documents we find it useful to describe the meaning of the term in the way that we use it.

The best definition of the term that I have come across is from John Seely-Brown the Chief Scientist of the Xerox Corporation and head of our renowned Palo Alto Research Centre. When told that he was to be the Chief Scientist responsible for the Research Division of "The Document Company" he admitted to being a little uncertain as to what was meant by the term.

A review of the dictionary elicited the following:

*V. Mital (ed.), Advanced Litigation Support & Document Imaging, 53–65.*
© 1995 *UNICOM Seminars. Printed in the Netherlands.*

- the term Document comes from the root latin verb Docere which means to teach. The suffix - ment usually denotes a concrete result of an action.

- an associated concept is the word Text, which comes from the latin verb Texere which means to weave.

From these two words John was able to develop his view of a document as an interpersonal communication protocol between people, an important mechanism by which people communicate to other networks of information woven from their learning experiences.

From this definition we can begin to see that a document can be regarded as a container for information, acting as a neutral means to present information to users, imposing few implicit rules on its creator and user by virtue of its role.

Taking this definition as a platform we can begin to see that a document will exist in numerous forms throughout its life. The prototype document will be a blank sheet, perhaps of paper. In this form it represents only the container itself. Make a note on a blank sheet of paper and instantly the sheet becomes a unique object which will communicate with any person equipped with the necessary analysis tool to interpret the symbols it contains.

As a single sheet of paper this document is uniquely useful to assist the human in the task of capturing thoughts and experiences as they happen for later analysis.

Later in its life the document may be discarded and the value of its contents lost, alternatively the information it carries may be transcribed into another prototype container which I will call the electronic document, you might know it better as a wordprocessor file. At this point the value of the original paper may be at an end as the "idea" it represented has been captured elsewhere, or it may still be needed in which case it must be carefully archived in a physical system where it will be able to be retrieved and used at a later date. By separating the concept of the document as a container for ideas or information from its content, which are the ideas themselves, and allowing the information to be represented to its users in ways other than on the physical paper medium, we can begin to understand how modern technology can be applied to improve the flow of information from its source to its user.

One of the approaches to the use of technology in this context is to create an electronic facsimile of the paper original which would ensure that the information contained in the original is preserved, while the actual original may be discarded, or archived securely to avoid the risk of its loss.

This electronic facsimile of the original is what we will call the "image" and this is the representation of the Document which is referred to in discussions of Document Image processing which we will be hearing about over the next two days.


## 7.2     IMAGE APPLICATIONS

When thinking about Document Image processing and how the technology might be applied to assist us in our daily work we, at Rank Xerox, like to think of the possible uses of this technology as lying somewhere along a continuum.

At one extreme we find the Transaction Document Image processing application, the easiest to visualise might be a customer account enquiry department of a major credit card company.

The document concerned is a transaction slip of which many hundreds of thousands arrive daily, the value of the transaction is captured, either at the retail shop or at a data preparation centre and entered into the customers statement.  In a minute percentage of cases an error may arise and a customer will contact the company to point out the error.

In order to process this kind of enquiry quickly the credit card company has invested in a large transaction document image processing system to enable the "image" of a transaction slip indexed by retailer and customer code to be found and displayed on a screen.  If necessary it will be printed and sent to the customer.

In this application we have a single format form which is stored by the many million to facilitiate rapid enquiry processing for a very small percentage of cases.

Another application in the transaction image processing end of the continuum will be the insurance or mortgage application case.  Here individual forms will be entered into an approval process and every form will be worked on by individuals as part of that process.  While the format is standard, each is unque, as it represents a particular instance of a person applying for insurance cover for a risk or for finance for a purchase.

In this case the company may have invested in "image processing" technology to support and improve the flow of work through the organisation.  Each document will pass through a variety of phases in the approval cycle and it has been found that it can be much more efficient to enable the flow by the use of electronic images of the forms which will move themselves from one phase to another in the computer supported world.

The paper equivalent of this process involves the physical transport of paper from department to department which, if analysed, may reveal that a particular case may take two weeks to pass through the process. Of that two weeks as little as 30 minutes might be spent actually working on the forms and the remainder "in Transit" from person to person or department to department.

The techniques to move the document through the computer supported process are called "workflow" and the design of workflow processes can, of themselves, be of enormous benefit in understanding the efficient processing of this kind of activity.

These applications are all at the cutting edge of "Customer Service" and "competitive edge" and are the current glamour end of the Document Image Processing Business. The benefits of improved access and controllable flow adds up to real and quantifiable benefits to the organisations using them.

At the other end of the continuum we find a different set of requirements which may be capable of benefiting from an "image" approach.

These applications address the fact that the traditional form of a document, paper has several important limitations:

•   as paper the document is inherently a single user presentation medium.

•   as paper it is uniquely capable of being transported in small quantities, and edited by its individual users.

•   as paper it is extremely bulky to store and when stored in bulk it is very difficult to retrieve a particular item from the mass without sophisticated, and expensive, storage and retrieval systems, probably better known to you as filing cabinets and filing clerks.

If we look back at our definition of the document as a container and presentation mechanism for information, we find, that while it does its job extremely well at the Human/Document interface, it is extraordinarily inefficient for archival and retrieval purposes in its paper form.

At this end of the document image continuum we begin to see the application of document imaging technologies as an archival and retrieval system which might enable us not only to store documents more cheaply and efficiently than we can with our cabinets and filing clerks, but also which might facilitate our ability to retrieve documents on the basis of the information they contain as well as any other characteristic which may be appropriate. This model of use is what we will call the "Reference" application.

We distinguish these applications from the transaction processing application set by examining the type of user associated with each. In the transaction case we see that the users of the system are participating in a process which has a relatively specific purpose. This might be to answer the customer query or to approve the loan application or insurance claim. In the main the primary document concerned is in a specific format, controlled by the organisation.

In the reference model there is no specific application or process and the users of the system are more likely to be professional knowledge workers, who are interested in the analysis of information and the actual use of the information contained within the "Information base". Here the documents will have no predetermined form or design and the sources will be both internal and external to the organisation, presenting particular problems for the archival strategy.

## 7.3    THE COST OF INFORMATION

This discussion brings us to a line of thought which we find helpful inside Xerox and which was developed by our research scientists at our Palo Alto Research Centre.

The concept is what we call the "cost of information" and closely allied to it is the concept of "Information as an asset".

We start from the premise that information has a value, and that much of the information that an individual requires is actually contained in documents. The problem is that there are so many documents around that bringing an individual person into contact with a particular document, containing the information he requires, at the instant that he requires it is an event of very low probability.

One can think of several special situations where by fixing one or more of the variables you can achieve great success, a roadside direction sign being one, but in the unstructured real world of information this is a difficult job.

The concept of the cost of information states that the easier it is for an individual to access a particular piece of information the lower is its cost, irrespective of its intrinsic value.

Thus a piece of information stored in the Library of Congress in Washington might have a high value but would also have a high cost as it would, in normal circumstances, be very time consuming and costly to go there to get it. But if I were able to make a simple enquiry on my computer terminal which would, through the use of modern technology, give me instant access to the database of the Library of Congress and all of the documents contained therein the "cost" of that information would be significantly reduced.

Similarly, one might think that the humble filing cabinet in the office would represent a low information cost as it is reasonably close at hand, but the accessibility of the information it contains is wholly dependent on the indexing system and the people associated with its storage.

If I have lost the indexing system or my secretary leaves, the cost to acquire the information in my own cabinet might reach the infinite.

I have in my own office a filing cabinet, much of the contents of which I inherited from the previous incumbent of my position, or rather the position I held when I inherited it. In the five years that I have had this cabinet in my office my role in the company has changed very significantly, in line with the evolution of the company itself. I have now reached the stage where I have a very limited idea what information is contained in this cabinet and I cannot afford the time to find out. It may be that there are people in the company who would find a use for the information in my cabinet but I have no way of finding out who they are and they have no way of knowing that I have it. The cost of the information that it contains is approaching the infinite and although its potential value might be high, the fact that it is inaccessible means that for all practical purposes its actual value is tending towards zero.

The primary cause of this increase in information cost in the paper model of information storage arises from the high rates of change in organisations today. If organisations were stable then the job roles and the information support to those roles would be well defined and controllable. But for good or ill the world today is not like that. Organisations are finding that the only effective survival strategy is constant change and evolution but an unforeseen price of change is risk that the people within the organisation become detached from the information required for their support.

What is happening is that the human indexing systems we use to maintain access to our files are unable to support the rapid changes in organisational approach and structures. If we lose the index it becomes impossible to access the information.

Does this matter? Should we be concerned about our ability to gain access to information contained in filing systems that were relevant to organisations and structures that no longer exist? Well the simple answer is that in many cases it will not matter. There is massive redundancy in the information that we store today, very frequently documents are filed for which there will be no further use, one might say that the vast majority of the documents filed today are in that category. The problem is that some of them will be of use, and all of them have potential to be of use. We must therefore define a strategy by which we are able to accommodate the storage of documents which hold the potential to be of use, in a form in which the future users of the information will be able to access them at the lowest possible cost.

## 7.4    INFORMATION AS AN ASSET

When we begin to apply quasi financial measures to the subject of information accessibililty it may be useful to extend the analogy to the status of information itself.    There can be no doubt that in every organisation the management of documents and the information they contain is a significant activity.    It is rare however to hear of organisations approaching documents as assets to be exploited.

Leaving aside for a moment the concept that information cost can be measured in terms of its ease of access and returning to our original view of documents as a concrete result of a learning experience, we see that the documents contained within an organisation represent in large part the only lasting record of the activities of the individuals employed by the organisation.    As we are all aware employing people is the single most expensive element of the cost base of an organisation. We employ people because of the contribution that they can make to the organisation's goals and as our organisations are more or less successful based on the results of their efforts, it would therefore seem to be reasonable that we should regard the collected results of their "learning experiences" as a very significant investment which should be exploited by the organisation to the fullest possible extent.

Accessing and exploiting the collected information asset of an organisation, often referrred to as the "institutional memory" these days, should therefore be something of a priority.    The fact that this has not been recognizably the case is more a function of the fact that new technologies to improve the management of this asset have only recently been brought forward, while the old ways of working, while inconvenient and cumbersome have not been seen as capable of improvement.

Today information in its broadest definition is increasingly seen as having value and organisations are examining how they can apply modern technology to facilitate its exploitation.

Because, in Rank Xerox, we are supported by a sophisticated document centric network, I can mitigate the high cost of information that is associated with radical change, by accessing electronically by means of electronic mail all of the people who may be able to assist me with my role.    In the main the response to a mail enquiry is dependent on the person at the other end being able and willing to both understand my enquiry and respond.    If I get a response it will normally be in the form of a document which may, or may not, be available electronically.    In either case it will be forwarded to me in an appropriate fashion.    The advantage of electronic mail in a global organisation is that I can send a question to a wide group of people many of whom will not be known to me.    Further, I can do this in my own time and it will be received and responded to by anyone who can help in theirs, their response reaching me when I next access the system.    While this system for reducing information cost is somewhat hit and miss it is certainly an improvement over the position that existed without it.

By the use of electronic mail we are able to access a wider pool of experience and knowledge than can be achieved through non-electronic systems, but this is by no means a perfect solution.  The imperfections are fairly obvious:

• there is a dependency on every one who can be of assistance bothering to participate

• because the respondents to electronic mail enquiries are humans, the only response I can get by this means will be from the current consciousness of the present participants.

It is entirely feasible that the information I require will be contained within a previous activity of the organisation of which the current participants may not be fully aware.  While electronic mail can, and does, assist with acquiring advice and experience from the current incumbents, it does not necessarily help in accessing the significantly larger "Institutional memory" which will be contained in the corporate archive.

If I were able to access by the use of technology the entire body of experience and knowledge on a particular matter, represented not only by present people but also by the document artifacts left by previous employees of the organisation, I would be better able to use today the significant information asset which is built up in an organisation throughout its existence.

In our search for a reduced cost of information we can identify any means whereby technology may be capable of assisting us, not only to gain better access to the talent and experience of an organisation today, but also to generate a return on the investments in information made in the past.

## 7.5     INFORMATION PULL NOT PUSH

Once we accept that information is an asset to be exploited, and that in the main this asset is packaged in containers that we call documents, we begin to consider how technology might be applied to enable this better exploitation.

In this process we arrive at a fundamental change, or paradigm shift, in the approach that we can take to the creation of documents and the process by which they are made available for use.

The present paradigm or the "analogue" model of document management focuses on the provision of documents in paper form only.  The document creation activity concentrates on the creation of a single master copy which is prepared by the information provider ready for "publication". Once the master is available, the publication process begins, often utilising the first significant document management technology invented by Chester Carlson some 50 years ago, the plain paper photocopier.

Our research in this, our heartland business, suggests that on average each page produced in an office will be copied some 19 times and circulated by the information provider to his target audience.

Notice that in this paradigm it is the information provider who decides who is to receive his product. The receiver of the information may review the document and then must decide if it is to be archived for future reference. On the assumption that it is archived, a proportion of the 19 unique copies will be filed in 19 filing cabinets and assuming that the filing and retrieval systems work correctly they will be capable of being retrieved and the content used when required.

The problems in this analogue paradigm begin to compound when the Document stimulates a response from one or more of the recipients. If the content is of value it should do so. The response may be in the form of a commentary back to the information provider and to one or more of the original recipients. As likely, elements of the document will begin to circulate among the contact groups associated with the 19 recipients and new knowledge and understandings drawn, all of these will circulate through the organisation in a haphazard and random fashion frequently resulting in new documents with different circulation lists and so forth.

While this method of document distribution works as far as it goes it has numerous weaknesses if we attempt to utilise the information asset that it represents.

Broadly speaking the basic problem is that of information "publishing" or "information push" as it has been described.

In the "information push" model the information provider decides when the information is provided and who receives the information, through its circulation list.

"information push" systems take little notice of the requirements of the potential users of the information. Firstly there is no guarantee that all individuals in the organisation who might have a need for the information will be on the providers circulation list — many may never receive it, or be aware of its existence.

Similarly there is a strong risk that, even if the potential user is on the circulation list, the information will not arrive at the point in time when the receiver needs it. Assuming that the recipient conscientiously reads and absorbs all of the documents that pass across his desk, it is quite likely that by the time the recipient actually needs the information he may have forgotten where it is or where it came from.

The new paradigm for document circulation attempts to resolve some of these problems by creating a new environment that we call "information pull". In an Information pull system the information provider creates his contribution and submits it to the organisations information repository, in addition to circulating it directly to the individuals that he feels may have an immediate use for it. The document is stored electronically on a system and all of the individuals within the organisation are supplied with electronic tools that enable them to enquire of the corporate information base about the existence of any documents which might be appropriate to the activity underway. The recipient and potential user of the information becomes any person, whether known or not to the information provider, whose enquiry matches to the index held for the document. The indexes may be extremely sophisticated going way beyond the simple indexing available on paper filing systems of subject, author and possibly date and extending to encompass all of the words contained in the documents and even synonyms and homophones and misspellings.

The benefit of the information pull paradigm extends beyond the wider availability of the information to all current participants in the organisation to all future participants as well.

More significantly, the potential user of the information is provided with that information at the time that it is required, which is when the user makes the enquiry and "pulls" the information from the document store rather than at the time when the information is first available or when it is "pushed" from the provider to the chosen audience.

This new paradigm of information pull supported by computer technology appears to address very well the problems and issues presented by our current processes for document and information management.


## 7.6     IMAGE APPLICATIONS IN THE LAW OFFICE

To address more specifically the legal marketplace we begin to find new complications which highlight requirements where the technologies may be deployed to gain an edge.

Legal practices and corporate legal departments can firstly be considered as organisations in their own right exhibiting many of the requirements of organisation that I have described.

But these organisations frequently are engaged in activities which further require them not only to have access to documents held within their own organisation but also to have access to Documents concerning matters which their clients have contracted for litigation. In these circumstances problems of the use of paper as an archive form for documents begin to have an added dimension of complication.

Firstly each paper archive now has two legitimate users, the organisation, who will be justified in attempting to continue its operations while a matter is ongoing, and the legal firm of advisers who will require access to the same documents for the purpose of preparing the case. The preparation of a case will involve a degree of turmoil for which the paper archives are ill designed opening up the possibility that important documents may be mislaid through accident or, on occasion, design.

In the preparation of a case important documents will be consulted frequently and will be passed around for opinions to be developed and the output of the opinions, almost always in the form of documents, must be circulated and managed efficiently.

The application of image management to this problem through the creation of facsimile images of the documents associated with a case enabling the document content to be analysed, reviewed and developed without undue disruption to the client organisation and minimising the risk that the documents concerned will be lost or misfiled can be seen to offer high potential benefits to the legal practice and its clients.

If this can be achieved, while also improving the ability of the practitioner to gain access to the documents and reducing the time required to retrieve the documents, it promises to revolutionise productivity and open many new possibilities.

We are beginning to see the use of document image and text retrieval systems in support of litigation in the US and in many celebrated cases here in the UK.

While we can see the stirrings of interest in this subject these systems have yet to achieve widespread use in the UK as to date there has been insufficient exposure of the possibilities of technology to the legal profession and of the potential of the Legal industry to the technology industries. Further the implications of the use of these technologies for the way in which cases are conducted and fees charged have yet to be worked out in detail.

## 7.7    THE INFORMATION USER

When we begin to discuss the application of new document technologies to specific problems it is imperative that we, as technologists recognise an aspect that is all too often forgotten in our enthusiasm, that aspect is the perspective of the user to the system.

In gaining benefits from the use of technology it is essential to recognise that these benefits only arise as a function of the human users' ability and willingness to use a system.  Unlike computer technology where matters generally conform to well known and predictable rules the human involved can be very difficult to please, while frequently the theoretical benefits of implementing a computer system are easily determined, the actual achievement of these benefits are more difficult to guarantee.

Frequently problems arise because the user community, prior to seeing the technological solution to their problems are unable to visualise the world as it will appear when the solution is installed.  Once installed the users are very quickly able to identify new possibilities and become dissatisfied with a solution that exactly met their requirements as originally specified.

The issue is of particular importance in discussing the potential implementation of "information pull" systems as the potential user is anyone in the organisation.  At least in the transaction processing applications we have a reasonably well defined problem to solve.

We often think of the application of technology to the office as having similarities to the experience of hill walking.  Ones objective is to arrive at the top of the hill but each summit we achieve only reveals the next summit and opens up all of the new possibilities that lie between us and the eventual goal.

It is for this reason that we welcome the initiative of Brunel University in organising this conference where these possibilities can begin to be explored and of the University in setting up a showcase of legal applications where the profession can visit and develop their understanding of what we as the Document Management Industry have on offer and can begin to explore with the industry the possibilities that lie ahead.

## 7.8    SUMMARY

I hope that in these few minutes I have been able to provide a platform and context which you will find useful in considering how the application of technology to document management may benefit your organisations.

I hope that I speak for my colleagues in the industry when I say that we welcome your interest and we hope that through this exposure and what may follow from it you will be able to formulate your own approach to the possibilities that technology may have in assisting you and your clients to access your information assets and exploiting them to gain that vital edge that will ensure your success in the future.

# 8. Understanding the Detailed Economics of Imaging

*John B. Massopust*
*Access Management Corporation*

The focus of this presentation will be on imaging as a component in the overall strategy and plan of litigation support. This presentation, while it includes technology, is driven principally by economics. Your primary focus, as you examine this technology, is framed in dollars and cents. As we proceed, this will become obvious. I would like to conduct a quick question and answer session to find out a little more about the audience. But rather than put you on the spot, it is a silent question and answer. As I pose these questions to you, place yourself in the appropriate category. We will deal with each of these and then move briskly through the information

The first step is some simple, straightforward definitions. I am going to offer two definitions that I think will stand us in good stead. The first one that I think is the best one I have seen in a long time, but it is a little long, so do not write it down. By the way, if you would like copies of the key overheads of this presentation, which was developed in Power Point, I will be glad to forward them to you, along with a couple of white papers.

Very good definition. However, there is a better one which I want you to remember. Here is how that works. Let us imagine for a second that this is the Kodak 9000 intergalactic speed copier. Two hundred bins faster than a blink.

This presentation departs from the presumption that you will automate a case. You will learn a method to obtain better control over the documents than simply putting them in notebooks and placing them on shelves. If we depart from that premise, this question comes into play: Do you need imaging? Yes, if two conditions are present. And I submit that the conditions are something like this. For those of you in the back I will read this. "Now that the desk looks okay, everything is squared away, yes squared away." The two conditions for both Mr. Einstein and the attorney are the same: (1) you simply cannot remember it all or (2) as in this case, your support systems let you down. We are going to introduce technology as a method of bridging those gaps, tying up those loose ends and

*V. Mital (ed.), Advanced Litigation Support & Document Imaging, 67–77.*
© 1995 *UNICOM Seminars. Printed in the Netherlands.*

delivering a superior product for accessing information at a cost less than you are currently paying.

You may be saying to yourself, or I would, "Whoa! I do not know if I am ready for this". Are you ready for this? Let me tell you a true story. A litigation support manager for a large firm opens an overnight Federal Express pouch and what does the person see? Five of these (optical disks) with a handwritten note on the stationery of the plaintiff's firm that is one time zone to the east, saying "any questions, give me a call". Sixty-five thousand document pages arrived in that overnight Federal Express pouch - "any questions give me a call".

Load the information, get the information. Did anybody go to Legal Tech? It is a technology show where most of the administrators for the large and medium size law firms visit rows and rows of booths and you have all seen those big screens, everything dashing and flashing. That is the output side that is 30% of the problem which commands 70% of the interest that deals with 30% of the costs.

As we examine the situation, it is the input side that is the most volatile and brings with it the real economic advantage because the retrieval side tends to work around the standard technology components the firm has. There is no real mystery to it.

The objective, very simply, is to get the right piece of paper in front of the right decision maker at the right time. The decision maker will vary. First, it will be the client. Then, it will be your partners, associates, legal assistants. Then, it could be co-counsel. Ultimately, it will be the judge and/or the jury.

It is a new way of doing things. There are certain things you really do not need to improve. For those of you in the back, the caption says "the modern lion". Now the lion has been doing just fine for a long time, but he would probably rest easier if he just could employ new technology. The economic agenda within which we must examine this will be from the law firm side. I recently had lunch with one of the more prominent law firm management consultants in the country. Shortly before our meeting, he had visited one of the largest retail organizations in the United States based in Chicago. That organization was very interested in applying for the Malcolm Baldbridge Award. The vice-president and chief counsel of the corporation called the consultant and said, "Could you please come in and talk to us about quality lawyering", about what it takes to do a good job because all the other departments - real estate, wholesale, distribution - they all have matrixes and checklists to define quality. We got him the next morning. What the heck is quality lawyering - budget, win/loss.

It was a really interesting discussion that continued through most of the morning. During a break, someone said, "I will tell you what it is not. It is not grabbing the leg of the Federal Express pickup person at 6.30 at night while you are waiting for the last page of the letter to pop out of the laser printer, hoping the commas are in the right place and a zero was not added inadvertently".

If, in fact, that is not quality lawyering, what would be? It is the time to honestly reflect and quickly make decisions on a substantial body of information. I would love to see a time sheet from an attorney that said, "thought about the case - 2.5 hours". It is not done. We leave piles around, a pile here, move this pile from here, add it to another pile and it grows. Move another pile and the piles continue to grow. It is similar to the lion marking his turf on the frontier. Suddenly, there are all these piles that are memorialized by line items on a bill. The essence of this technology is to reflect on the information provided by the imaging system and the automated litigation support system.

More than anything else I have seen and I have been doing this for a long time, imaging is a piece of technology that fits perfectly. This technology helps to eliminate some of the details that are essential but are annoying to corporate counsel managing litigation. We had such a corporate counsel in our offices some months ago. He is responsible for managing approximately 100 out-of-house counsels across the United States for a large corporation in the petro-chemical business and he calls this stuff "fog". It is all that stuff that you would call variable expense associated with lawyering time dealing with the management of information which is necessary to be effective in litigation. If you address this, technology will help you leverage it.

Here is a definition that I purposely withheld and am now putting it in this section. Major litigation - less than half a million pages. Do you have a room like this? Have you seen a room like this? have you spent any time in a room like this? Well, one of the things the imaging system does is that it gets rid of the boxes. For example, here are 150,000 pages right here, available instantly, a button push and it is on your screen. That is the impact - less than an 1/8" high, seven gigabytes of data; capture, unload, retrieve.

Here is the problem and also the purpose of my presentation: it's about paper. You desperately want the information and you want to find those key documents. It accidentally happens to be on paper which, by the way, has served us very well since 700 B.C. With a minor improvement just before World War II in microfilm, it has been paper all the way. Big change, big impact, economic reality, technology options, right today in use.

Here is a flow diagram of what happens to paper in major litigation as it is coded. Starting on the left with the large document population, it enters the system - the little green circle over there says screen. The total universe is then refined to a document population that is prepared for treatment. Then it is treated and goes to the photocopy machine.

We interviewed a large pharmaceutical company that was compelled to use imaging to support their litigation because, and I found this hard to believe and had to ask the chief counsel twice, the paper was wearing out. It had been stamped so many times that the stamps could not be tracked. It was wearing out and every time it went through the photocopy machine, it cost $0.15. If you are ever in a room where that "go copy it" comment is made, stop for a second.

Once you pivot at the photocopy machine, much of the compelling economic advantage for imaging is lost because your spreadsheet of what it will cost from that day forward is burdened by a reproduction bill for a couple hundred thousand pages. Stop right there - do not pivot. Let us look at it.

You will notice that the copy box now is split. You either copy or scan. But it is either this or it is this. Due to information needed, in this case, my marginal notes, my voice over for this script is literally more important than this. I need the data from it and it accidently happens to be in paper. Scan and then someone looks at the image. Frequently, I am asked, "But if it is so automated, then it is so compellingly cheap and the people are not in it and the quality suffers". Every image is observed by a person. Can you see where it says "QC or Flag?" QC means it is okay — not smudged, did not move, not sideways, not backwards. Since it is okay, the picture took. If it is not, it goes in the "to be redone" pile to go through the process again and, ultimately, you will get a good picture.

Flagging are electronic "stickys", the little yellow stickys. This came from the Rocky Flat site and was generated before 1985. This was on microfiche. I can add a variety of yes/no or quick, single item flags to the electronic image instantly during the capture process, very inexpensively. It is a very compelling argument to scan to build the archive because as the archive is built, let us think for a second what 100,000 pieces of paper copied will look like. Two-wheelers bring dozens and dozens of boxes in and there they are. As opposed to loading it onto a system where they are already numbered and where you could segregate them by flags — these are old, these are privileged, these are ours, these are yours — all these categories can be indicated by flags. That happens at a flagging station during the capture process. At this point, I have two options. I can use that electronic file to index every word using OCR technology - we will talk about that in a second or I can stop there and simply generate the image database.

Blue circle on the right (the end user database). That is where we are trying to get to. That blue circle consists of two elements: the text portion and the picture. The text portion describes the document, either full text or certain coded information. You can expect to get the blue circle, the user database, any way you want it. I want this in a certain format because it runs on my machine. Okay, I can do that. I also want it to run in any one of a dozen different text engines that I use all the time. As a matter of fact, I want it two ways. I want it in this one and this one. Okay. Those are just formatting questions. They do not have anything to do with the data or the coded information. The good news is that there are a variety of options in delivering pictures and text to firms that need to build large databases and you can simply specify what you want. Knowing what you want is where we are now.

Input/Output is a simple, retrieval strategy. It is the key to the success of all of this. What are you looking for? Well, if you are in the middle of the woods, that could be your target. Would you rather look at that or would you rather look at that? I feel pretty certain I could hit that bulls-eye which is a comfortable transition into our next segment

OCR — it is the ultimate index. It is every word on every page retrievability. I would like to emphasize the first bullet. The current state of technology is **not** what I remember and you remember from the early and mid-80's. If you scan, if you create an image, you now have an electronic file of every character, every letter in every word on every page. I pass it down a network pipeline. I reach an optical character recognition machine and it indexes every word - 600 pages an hour, 700, 800. Depending upon the quality, over 800 pages an hour, fully indexed, which means fully retrievable, ready to go. It is very compelling.

Here is a document. It is the first page of an environmental engineering report. I know it is real because there are already two numbers on the lower right-hand corner. This one has been passed around. Here is the original. Now, here is that document with five bibliographic fields so you can see what is not there. This is with five. This represents some money, these five fields will cost you something and so there is a budget. Here, it is with eight fields. You can see that I have added the title of the document or the re: line and from the back of the room, I can tell you it is three and a half long lines so it is a pretty good task to put it in there. Names mentioned, every name, every organization mentioned.

You make your OCR decision when you understand what you are trying to find and what you are trying to find it in. Most of the documents that I have seen in cases are pretty bad. It is not hot off the HP Laserjet and right into the system but for second, third, even fourth generation documents, we are constantly amazed at the quality and the high percentage of accuracy that we can achieve with high-speed OCR. I am frequently asked what is the accuracy of OCR? What do you need to have? 100%. It did not take you long to say that. 100% or not interested but you could be interested.

If the title is mentioned and organizations mentioned, it will cost you $0.60 per page to get. You might not have the budget to do it. But you may have $0.20,

$0.25 for this task and you are talking pennies here per page to OCR the whole thing. The percentage that you miss is a value judgment you make based on the available budget to retrieve the information. There are systems that can report how accurate they are.

As this page goes through and it is a typed page, the system will instantly calculate I read every word, every letter, every character I substituted, no question marks or asterisks, I get 100%. The machine tells me that. Because if it cannot read it, it inserts a question mark. You can set the threshold that makes you comfortable. I have to achieve 80% at the machine level. If not, it is placed in the "to be addressed later" category. Therefore, I can grind through half a million pages, make a value judgement on the quality of indexing and then decide if I have discretionary money available for additional indexing.

Let us talk about that page. There are three components: (1) on the page, (2) the page and (3) about the page. The document and the image, paper or an image format in the electronic file. Here is how I tie those together. As a single package, this is what sits in the blue, the database, the client database which is a combination of words and pictures, text and scanned information. It is a package. It is as if this picture and the words associated with it have header information that tie the whole thing together in a retrievable package.

You start up your search engine and ask it to find all the documents where Harris and Simpson appear in the same paragraph with Minnesota in April. Four documents meet the criteria. I would like to see them. Push a button. There is the first one. Pretty strong. All of this is on the input side — building the database, making judgements, establishing your budget.

Also, there is software; two types — the software that manages the images and text software. First, the imaging software. What does it do? It takes the picture. Once it takes the picture, it compresses it for storage and makes it retrievable. No more paper. Mission critical applications supported by imaging. It is an element that must be evaluated in paper intensive litigations.

If this is the picture, the header information sits right here and actually talks about the image itself. By the way, it is usually proprietary. I use the "P" word with a lot of confidence now. In the early 80's if you used it, it was death because we were talking about word processing and you remember those dedicated word processing systems that could not talk to that one. As a result, we had the "W" system, the "I" system, the "N" system, the "S" system and everybody said, "I do not like this". PCs came along and everybody could talk everything. Do you ever wonder where all those executives went from those word processing companies that sold proprietary systems? They went to work for imaging companies because it was the next technology. They are there now and there is something very good about the proprietary nature of the header records that describe the images here; that is, all of those proprietary systems can get to one or two basic conversion definitions.

I can, in fact, without knowing anything, get somebody's file format, load a couple hundred thousand images and a month later decide I am going to change imaging systems and convert those to something else. I would not know a TIFF if I saw one but you can always get the TIFF format to convert or PDA or other acronyms that really are not important. The good news is that you can move from system to system far more economically and far more easily than you could in the past.

The retrieval software is at play now and we can use it. We load information in big text databases. As a matter of fact, we had a group in our office yesterday and we put 27 candidates up on a piece of paper, starting with the B's and going through the Z's and there were 27 that are used. There are many choices available from very expensive because they require big computers to what we can buy at retail shops on the way home today. You can have anything you want. I suspect that within the next year or so, there will be multiple engines software actually used on the same database for searching. Some systems handle text very well but are not effective with fields. You should have a good relational database for the fields, but full-text indexing and relational are at the opposite end of the mathematical formula for indexing and retrieval. You would probably use both. Windows make all of this possible. You will actually have the text engine searching on the left-hand side of the screen. Hit F10, F7, F-something. Up pops the full size, regular picture of that page right on the screen and you can actually type notes in a third window. That does not impact the firm strategy for software platforms. You can introduce several Windows applications for a specific solution simply by running the application in Windows, and when you are not using that application, you do not have to run the Windows which provides more flexibility.

The output. The strategy and the hardware. Oh, the strategy. All the gray hair I have now resulted from the past nine months I have spent explaining to people how the economics work. Because the problem we have is — when do you start, how do you start? The case — lots of paper, good client, let us image it. That is a wonderful first thought and I hope you do it. Remember the photocopy machine, the one that is in your office now, the biggest one you have. Would you buy that photocopy machine and assume that a single client would pay the freight on that entire photocopy machine? The photocopy machine is a tool as well as the image retrieval system. The math may work that the retrieval strategy, the number of documents, the number of co-counsel, the whatever all work so that the case pays for everything. Input, output, database building, the whole nine yards. It may well be the math work or the spreadsheet will not quite carry it all and, consequently, I do not want to image it. Well, if you do not image it, what I submit happens is that it comes right around the barn in the "stuff" category or the "fog" category. I suggest that the opportunity to "stuff" and "fog" will be under very close examination as we proceed with information management solutions.

In a perfect world, we would say that, today, imaging is with us. The technology is mature, I understand it and I understand that the large document case that I have today is at play now and will not be the last. Will that be the last major case you have? No. For the next one, does the unit cost go down as you amortize the expense of retrieval stations, other connections and bits and bites of technology? Certainly it does. It is comparable to the cost per page for your big photocopy machine. The strategy is everything. Not only going in when you are trying to decide what you need, but coming out the other side. What is it going to cost to get it out?

If you had the retrieval capability in-house now, the math works great, the speed is compelling, the accuracy far better than we can get anywhere else. It is a simple decision. Sometimes the decision is taken away from you. For example, a large Fortune 100 company in the United States with close, ongoing relationships with out-of-house counsel handling their litigation work for years decided, within the past 18 months, they were simply going to take care of the variable expense of information management, a/k/a producing documents, coding, indexing. They are going to do it. They bought an imaging system and did it in-house - coded, loaded, took pictures, and everything. What do you suppose they did? They bought another system, loaded the pictures and gave it to the law firm. Here is your database.

They just took an unknown called the variable expense associated with producing and managing the paper and defined it. They did not just define it; they put it there and said, "done". They want the lawyer hours. They want the attorneys. They want the culture. They want the expertise. Of all the corporate counsel that I have had the opportunity to interview and discuss this matter with, they all say the attorney hours are not in question. It is the "stuff" that comes along with it that is going to stop.

The hardware elements need to be discussed briefly and so just a quick pass through here because I assure you, this is not the challenging part of the equation. There are only three parts. The PC, nothing complicated here. For the Apple users in the audience, Apple works fine. I am not that familiar with Apple. Big screen. Absolutely. Why? Because this piece of paper will appear full size, crease for crease, smudge for smudge, letter for letter on the screen. Will my color VGA work that I currently use for word processing? Yes. Works fine. But do not use it. About $1800 for this. It is the key part of the system. Storage - 5 1/4 MO (magneto optic). Erasable. I can write over it. It is like a disk. Scratch pad. On the other end, my good friends at FileNet let me carry this and if I ever broke it, they would be upset. Seven gigabytes, 150,000 pages. This is now the state-of-the-art.

Here is what it looks like as a stand alone station in a firm. Simple PC, a scanner, low-speed, nothing very big and expensive, a printer (your printer), and probably just a disk drive that handles these because every time I move one of these, I move 20 banker's boxes. Bigger means more retrieval stations. There is usually a PC server that manages the tasks. The storage over on the left is probably an autochanger also known as a jukebox where you have got these all stacked up and ready to go. Here is what it looks like in a big situation. If it is a seven or eight floor law firm, you just hook into the law firm's network and there are servers that manage the individual tasks. They manage the input tasks because there will probably be lots of input. Lots of retrieval that is managed. Output - printing and retrieval. But all the pieces stay exactly the same. Nothing very complicated. You tie all this together with "green stuff" which I call processing software. That is the difference between the entry level systems where you look at things and where you actually package tasks together with a single click.

For example, as you acquire the input from all these various tasks and you bundle them together with this "green stuff", this hooks in software and ties it altogether. For example, big database getting ready for a deposition; create a depo pack. Okay. John is the author. I want to see everything that John wrote and everything that he was a "cc" on, etc. I need it in date order but I do not want to print it all, just the first page. I want to see the first page to determine if I want it. A single button click. Complete scan through the database. Sort by date. Pivot on John. Print the first page only. Here is the attorney. Yes-no, yes-no. Take the yeses back to the screen, bar down, print-print-print, cue, print, there is the package.

My last riveting, technical contribution to this presentation concerns input resolution which is the speed with which the scanner takes the picture and is also known as dot density. Have you ever had free time and actually stared at the dots that create your fax ? Have you really ever looked at those? Here is what dots mean. As the scanner takes the picture, if the scanner is taking at 300 dpi, I can OCR from that file. Did you all get that? This serves as the input file for indexing. If my input speed is 200 dpi, the dot density is less than 300, I cannot OCR it or my OCR opportunity may not be as good as it should be. 300 dpi is a good note to take in this presentation.

The compelling numbers. It is true that litigation meets all of these criteria. And we are also in agreement, I believe, that it brings with it - this. Comparable to a tail on a kite. I am more interested in the kite. I have to include the tail sometimes, but now I have an option. You pivoted at the photocopy machine. So what. Real numbers. This is a million pages. The green line on the X represents the costs and I am estimating that I can do that for $0.11 per page to copy 15 sets of million pages. As opposed to imaging, first copy is $0.15, second through 15 is $0.025. What is all that stuff in the middle? Let us role play for just a second. You are corporate counsel responsible for managing worldwide litigation for a large New

York based technology company. You are hotshot litigation attorneys with lots of experience in managing complex, document extensive litigation. What is that stuff in the middle? Meet outside and discuss that. Okay.

Hard numbers. We just looked at duplication. Now let us look at the impact of OCR on that scanned image. Bibliographic, say five to nine fields, in-text coding, the example was names mentioned, organzations mentioned, what is the other one mentioned? There was another one mentiobned. Yes. OCR. Let us change the role play. You are the corporate counsel, you are the litigators, meet outside afterwards, talk about the quarter over a million pages. Say it is a million page project. Twenty-five cents look a lot differently over a million pages.

More hard numbers. The big red chunk "labor" is filing maintenance and retrieval. 70% of the cost of maintaining paper-based systems is in the labor. As opposed to image-based systems, the labor component shrinks by a full third. What is "X"? We have the opportunity to audit a very large piece of litigation. This is the profile and these are our results. We estimated after we did a line item examination of literally 100 pages of law firm invoices to the client and looked at everything from every disbursement charge, to every cab, to every bellman, tip, taking depositions during that period of time, that approximately 70% of the bill was lawyering. It was hours. It was lawyering. On the right 30%, the information management component was the stuff that got "foggy".

On the other side, there is no corresponding strategy for information management. That is the preemptive move that we are observing that corporate counsel are now taking. I want to control that because I need the hours and the hours are the observable execution of the litigation strategy but on the other side, my law firm needs help. I am going to give that strategy a push. I am going to execute that strategy.

This presentation always gets back to money. If you do not image, you sell the client a $45,000 copying bill and associated labor cost, so the "X" factor of savings is never at play. However, if you redefine the process and envision what it would be like to have all those documents in this particular case, 12,000 of them, then you can bring it up on the screen in a blink and print in another blink, as opposed to "can you cover the phone, I have got to go pull some documents this afternoon". You go to the room and pull documents and you see what the room looks like.

Imaging will have a major impact on how you conduct information management. The process does change or as this picture suggests, as if we all know where we are going, you can just keep on copying, filling notebooks, doing what you are doing until someone says we should stop.

A couple of quick concluding notes on the benefits that are pretty obvious. I would like to state them for you again. Doing business differently, like having a LaserJet near your desk. What a difference. Do you remember what it was like before you had a LaserJet near your desk? Try hard to remember. It was less than 24 months ago I submit for most of you. Doing business differently. What happens is a change in the process. You do business differently now. If there is one thing that corporate counsel understands, it is the key word - quality.

# 9.   Managing the Firm's Knowhow

*Peter Ilion*
*Beachcroft Stanley*

## 9.1   WASHING AND IRONING

Friends of Information Technology ought, I believe, constantly to bear in mind what that great lawyer, Lord Mansfield, said of the action for money had and received. "I am a great friend to it", he said, "and therefore I am not for stretching it lest I break it".

When I was a child I lived in what is now called Zambia and naturally in those bad old colonial days we had several servants.  One of them, James, had been with us many years and for most of that time he did the laundry in the only way it could then be done.  He laboured long hours over a tub of water, washing and scrubbing, and then he ironed.  One day my father obtained one of the first automatic washing machines in those parts.  James watched its installation with intense interest and when it was over Father asked him what he thought of it.  James hesitated for a moment or two, and then, clapping his hands in the traditional show of respect, said in a tone of deep disappointment: "But, Bwana, it doesn't do the ironing".

The runaway success of the electronic digital computer, and the clever ways in which it can be programmed to mimic activities which once seemed to require "intelligence" tend to raise expectations far above anything which the technology delivers.  Whatever we may one day discover about the ways in which our minds work, it is clear that the equipment and applications available now at a cost economic for our practices necessitate a very considerable amount of human help. The need for highly skilled technicians and craftsmen - engineers, programmers, system analysts and the like - is taken as read.  What is consistently under-estimated is the effort and time users have to put in in order to learn and understand how to use the facilities given to them.  The machine can do the washing but we have to do the ironing.

Whilst I realise that it is a currently most unfashionable view, I strongly believe that machine is best thought of as a tool - a very powerful tool for the mind — but a general purpose tool nevertheless: something like a knife (a Swiss knife perhaps) rather than a platform on which to run idiot proof applications which provide the

*V. Mital (ed.), Advanced Litigation Support & Document Imaging, 79–91.*

Meaning of Life, the Universe and Everything at the push of a few buttons.

## 9.2    KINDS OF KNOWHOW

I find it convenient to think about knowhow according to the manner in which it comes into a typical system in a legal office.  Three broad categories are:

- external
- specially generated
- automatic.

External data is stuff like Lexis and CHORUS, accessed through modems, and bought in machine readable data such as Justis CDRoms and Butterworth Precedents on floppy disks.  Automatic data comprises letters and documents and the like which are generated automatically in the sense that they are on the system as a by-product of the daily wordprocessing.  By specially generated data I mean things like precedents, counsels' opinions, internal manuals, telephone lists and marketing data bases.

The shape of external data is predetermined by its suppliers:  there is not much one can do about it other than try to make it as conveniently available to users as possible.  There is no technical reason why it cannot be made continuously available to all users at their desktops.  Unfortunately two factors virtually cancel out the whole potential benefit: cost and obscurity.  The prices of the CDRom versions of legal and other material is typically two or three times the prices you pay at the bookstore;  and even then there is the typical chant: "Single User Licence for only One Machine".  What makes the publishers think that we would ever want to buy 150 copies of their highly-reputed Treatise on the Law of Grommets?  One or two copies in the library was quite sufficient.

The suppliers of Lexis refused to let us put it on our network for "marketing reasons".  Actually they did us a service because by effectively causing it to be made available only in our Library means that at least our Librarians have the opportunity to use it often enough to learn how to use it.  The Justis EEC Law CDRoms are relatively easy to use and there is the added comfort of not incurring online charges: but access is still buried under the software they provide.  Each supplier of this sort of data seems to expect us to learn their special language.  Imagine having to know a different dialect for every book you pull off the library shelves.

I could go on and on riding this particular hobby horse of mine but instead ask you to contrast the £2000 BT charges annually for its CDRom version of the telephone with the French who provide not merely free access to this information but free terminals  to do so.

Incidentally the London Branch of the Society for Computers and Law did investigate a very promising project for the creation of a generalised access system for different external data bases using a single set of commands, but it floundered — for reasons of cost.

## 9.3 VIRTUAL MACHINE

This arrangement has also enabled us to set up what, to the user, is effectively one single large virtual machine and at the same time distribute the main processing load, wordprocessing, onto the users' PCs which run standalone copies of WordPerfect. Documents are printed by shared printers, managed by the UNIX print spoolers, and for letters, secretaries have each the sole use of a printer, or share one printer with not more than one other secretary, directly connected to their PCs.

## 9.4 FILE SYSTEM

The key to the virtual machine, and to the management of our information and knowhow, is the file system. Because the word can have several different meanings I should make clear that in this context I mean "file" in the operating system sense of the basic unit in which data can be stored on the system.

Like most legal offices as a new matter comes into the firm it is assigned a unique number for accounting and file puposes - the matter number. On our system the first thing to do when a new matter comes in is to open it on the computer. The system makes the necessary entries on the accounts system and, on one of the other servers, opens a directory with the same name as the matter number. In order to distribute files reasonably evenly across the various file servers subdirectories for departments are set up on different servers. Each departmental directory has a subdirectory, called "matters" which has a subdirectory for each partner in that department, called by that partner's official initials - the matter partner directory. The matter partner directory contains all the matter number directories for the matters for which the partner is the "matter partner". Each user has a home directory within the departmental subdirectory. There are therefore at least two directories named by the partner's initials but there is no confusion as they have different path names. All matter related work goes into the matter number directory, whoever creates it on whatever machine in the system. There are also departmental and global subdirectories for precedents and, in due course, for counsels' opinions and other kinds of specially generated knowhow. Our staff are free to make as many subdirectories within the matter number directory as they please — indeed we encourage them to break up their work in that way — as long as they all hang under the correct matter number directory.

By the use of symbolic links, and NFS cross-mounting of files on other servers, the file system is made to look like a single directory, rooted in — what else — a directory called "BS". We also use symbolic links to avoid another potential problem with deeply nested hierarchical file systems — long path names. To go to the right matter number directory all you have to do in UNIX is enter /m/12345 or in DOS M:\m\12345.

The way files are stored therefore closely resembles the way in which paper files were, and still are, physically distributed in our office: and users understand it as soon as they understand the simple concept of a tree based (hierarchical) directory system. That concept is a most powerful paradigm which should be fully exploited in organising your data. One of Life's Great Mysteries to me is why users who happily navigate the intricacies of wordprocessing systems and spreadsheets roll their eyes in horror when it is suggested that they might use the file system at operating system level. Sadly this attitude is encouraged by system administrators who may perhaps be forgiven for fearing the Great Dragon, rm*, who with a single breath can wipe out the entire system if it is not properly administered with appropriate file permissions. But the attitude reaches up to the highest levels of system design. One of the reasons why a famous supplier did not get our contract was that their designers had decided, in their wisdom, that four layers of directory were the most that any punter could have.

## 9.5    FINDING INFORMATION

The file organisation described also makes it easy to find files if you know more or less what you're looking for and are able to associate it with the matter number or with enough matter header information to ascertain the matter number. But what if you don't know enough to find the right matter number directory or are unlucky enough to end up in one of those matters with hundreds of files? Looking through file names is not much help for the reason I have already mentioned.

To take a concrete example: you are writing to a client about a particular legal subject and want to know what, if anything, your firm has already said on that subject. This is not only because it may be uneconomic to repeat the research and thought involved but because you would like to prevent different members of the firm giving conflicting advice to the same client. Yes, it does happen.

This brings me to what I believe to be central to the business of managing knowhow and other information on computers. Computers, at least as they are currently used, do not understand, and cannot understand anything. They are simply token manipulators and counters. This is not in any way to denigrate them or to detract from their immense power, nor to suggest that I necessarily think that they will forever be incapable of understanding. It is merely to describe the existing state of affairs with regard to the machines and programs currently available to us for our offices.

Naturally, there is a strong connection between ideas and the words (ie tokens) used to express them. So searching for these words and related words and combinations of them will take you a very long way to finding what you want but it will not necessarily take you all the way. The more you help the computer the more it will help you. The way you help it is by leaving tokens around for it to find.

## 9.6    DOCUMENTS AND LETTERS

This is particularly important in the case of legal documents which are characterised by their relatively restricted vocabulary and the high degree of redundancy in their wording: you would find it difficult to distinguish the relevance of one average commercial lease compared to six others merely by comparing the words they use. The words which do differ amongst them are likely to be the least interesting: names and addresses of parties, descriptions of properties and the like. That sort of document should therefore as a matter of course include a paragraph or two of key word information: the sort of description you would give if a colleague asked you to describe whether or not it contains anything of interest and it's just as important to say that it hasn't as to say that it has. If you can do this systematically so much the better.

It is too much to expect your fee earners to do something similar when writing letters of advice. Luckily letters of that kind contain much more information: you would be hard put to write about, say, section 140 of the Companies Act without using those words in close proximity to one another. But even here fee earners can help the machine by, for example, sticking to the firms standard reference format and resisting the macho thrill of using their very own exciting personalised version. The reference can provide useful classifying information. It is a good practice to start all advice letters by stating as clearly as you can what exactly you think the client has asked you to advise about. All too often in my career as a solicitor I have had occasion to remember the remark of a wise old principal I once worked for: "Peter", he said, "when lawyers find themselves in difficulties about how to advise their clients, it is usually because they haven't been properly instructed. Make sure you understand what your client thinks he has asked you to do".

## 9.7    ENTROPY

Because it is so important to understand that the computer is only an inanimate tool, and that what it produces is the work of its wielders - who would give credit for Michaelangelo's *Pietá* to his chisel? — I would like to quote what Richard

Hamming says in his book, **Coding and Information Theory** about information content. He is the discoverer, amongst other things, of the error correcting code which bears his name - error correcting codes are, surely, amongst the most beautiful discoveries of our time. He explains that the information contained in a message or other text is measured by the amount of uncertainty in it: if the message you receive is exactly what you expect, you gain no information from receiving it. He then says:

"This is an engineering definition based on probabilities and is not a definition based on the meaning of the symbols to a human receiver. The confusion at this point has been very great for outsiders who glance at information theory; they fail to grasp that this is a highly technical definition which captures only part of the richness of the usual idea of information.

To see how far this definition differs from "common sense" consider the question: What book contains the most information?............................ the answer is clearly: The book with the most information is the one with the type chosen completely and uniformly at random! Each new symbol will come as a complete surprise."

The meaning passes from the human inputer to the human recipient; the computer is only a filter. Do not have too high expectations of fancy induction engines based on this mathematical idea of "entropy" or of "fuzzy logic" a kind of probablistic logic: both very useful tools — in the hands of a skilled user.


## 9.8    TEXT RETRIEVAL

If you are dealing with a few hundred files kept together in one directory and are looking for words or parts of words, you can use grep, the UNIX regular expression finder. Most people are familiar with the idea of global searching with "wild cards" in a wordprocessor which is a simplified form of regular expression search. But if you must look through gigabytes of data text searching directly on the text is far too slow, and too greedy of system resources in a multi-user system. Nor are the patterns of grep to everyone's taste.

The standard solution is to use a utility which makes an index of all the words in all the files you want to search over along with their addresses on the computer. The index is kept sorted so that the software can quickly find the words you are looking for in much the same way as you look up a telephone number in the telephone directory. Imagine having to find a London telephone number by starting at Watford and asking at every door as you wend your way south until you find the right number. In other words, these utilities give you effectively random access to the words in your files in much the same way as your operating system gives you random access to your disk files.

An important difference from the management point of view is that the disk operating system makes and amends the index (i.e. directory at the high level — file allocation table at a lower level) transparently every time you create, amend or delete a file, the text indexing utility has to be told to make the index. So there is a potential problem on a dynamically changing set of data like your letters: the index will be out of date for new or amended files until the index is updated. Some kinds of text retrieval software, (BRS for example), try to resolve this by keeping a compressed copy of your files and it is the compressed data which is indexed and retrieved. This obviously makes indexing slower and while it solves the problem of the index pointing to the wrong places, it does not solve the problem of the data you retrieve not being a true copy of what is now on the system. For various reasons, I prefer the approach of the system we use, Topic, which indexes the files themselves. At least if the wrong words are retrieved you know that the file has recently been changed. It doesn't look as if, in practice, this is going to cause much trouble — indexing on Topic is fast and updating is even faster. So you can do it frequently. I would expect that the next generation of software will make these indexes for you transparently, just like a disk operating system and now that some operating systems are starting to let you keep all your files in compressed form gaining impressive reductions in the space they use up, typically between 30 and 50 per cent for text data, I suspect that the Topic paradigm will win.

One other problem with these indexing systems is that wordprocessing systems keep their formatting information embedded in, and scattered around, the document itself. So there is a lot of "noise" which the indexing and retrieval software has to cope with. Wouldn't it be nice if wordprocessing programs kept all such information collected together at the top or bottom of the document, or even in a separate file, together with pointers to the places in the text which are affected? This must now be technically feasible given the speed of current and soon to be expected PCs.

Topic has appropriate filters for the commonly used wordprocessing programs, but you have to index each type of file separately. This is transparent to users but does complicate the administrator's task. We had hoped to use the filename extension to distinguish pure text files from WordPerfect files, but that hasn't worked because users will not be meticulous in using the correct extension. Why should they? They have better things to do. Luckily, WordPerfect files all start with a "magic" number, and it is easy to include it in the file of "magic" numbers which the UNIX command find and other commands use to identify what kind of file a file is.

Topic has a number of other goodies. For example, it can with be set up so that you can keep profiles for particular users containing the "topics" in which the user is particularly interested, and when something arrives on the system about that topic it will be mail-boxed to him.

## 9.9    HIGHER LEVEL SEARCHING

You can rank ways of finding documents as follows:

* by searching for strings (i.e. sequences of characters)
* by searching for string patterns (such as strings bounded by non-alphabetic characters, e.g. words)
* by searching for boolean and proximity combinations of strings
* by searching according to measure of the occurences of strings within the document
* by using rules based on combinations of these methods.

I do not pretend to have mastered all the intricacies of these methods nor will bore you by showing off what I do know. But there seem to me to be two basic approaches:  you can learn how to make effective use of some or all of these methods yourself which is not that easy as they are not particularly close to the way we think, or do what most busy practioners usually want to do — you can let someone else do it for you.

I feel uncomfortable about the increasing trend to delegate legal research to librarians, paralegals, trainee solicitors and the like. So much is lost in the transmission of the message from principal to delegate, and you lose the serendipity you have when you are working in a well organised library.

Topics in Topic are an attempt to let you have a foot in both worlds. They are essentially searching rules in the sense I have mentioned which can relatively easily be put together and used as if you were searching for words. So you can get your specialists to create various topics for your firm which can then be used by your fee-earners. For various reasons we have not yet had a chance to go into these vary deeply but I believe they represent a very promising approach. Nevertheless you have to bear in mind that it is still doing nothing except token finding and you can be misled if you don't understand what it is actually looking for.

## 9.10    FREE TEXT Vs STRUCTURED DATABASES

Topic has another feature, which they call "filters", and which enables you to emulate the fields of a conventional structured database and, in a sense, to use it like a database. For example, finding matter numbers on our accounts system is a doddle if you know the client and there are only a few matters current for him. If, as we have, you have clients with several thousand current matters you have to think about narrowing the search based on the contents of the matter description field and you have a real problem. Like most structured data bases the system is optimised for fast retrieval of field based information and its string searching facilities through numerous instances of the same field are primitive, to say the least. As the matters are opened, our system, in addition to making the appropriate matter number directory on the right machine together with the symbolic links which make it easy to be accessed by users even on other machines, also copies the matter header information to another place where we index it under Topic and the information in it can be almost instantly retrieved using the powerful facilities of Topic. In this case, a particular fee-earner's initials may appear on different matters in different capacities — client partner, matter partner or matter fee-earner — so you need some field structure and the filters of Topic let you do this. A similar approach could be adopted with any text retrieval system which does not have filters but does have Boolean facilities.

Our use of Topic in this way does infringe one of the cardinal rules of data base management — not to keep more than one copy of the data (except for backup purposes) — but it is forced on us by the nature of the accounting software. Topic does allow you to import and export data between a number of database management systems, Oracle and Ingres for example, but this is not available for the Sculptor interpreter under which our accounts run. Hopefully, our next accounting package will remedy this.

This leads to a major point: there is quite a lot of information in the accounts system which is useful for the general office and for things like client and marketing databases: and it really should be available for document processing as readily as it is available for accounts purposes. This is not all that difficult if your accounts system is built on top of a database manager which allows "dynamic data exchange" with Windows and your word processing does the same, as Wordperfect for Windows does.

There are three difficulties with using a structured data base outside the accounts and administrative areas:

• your fee-earners have to learn how to use it
• they have to have the discipline to keep it up-to-date
• the data base has to designed.

This last is a subtle point: you have to have a fair idea of the kind of information you wish to extract right at the initial design stage. Not only does this delay implementation of the project, and greatly increase its cost, but you are stuck with its design until you can start all over again. This is just feasible for administration-type applications but it does not seem appropriate to me for knowhow management. To me the great benefit of a computer based system is not in retrieving information you plan to extract — manual systems don't do too badly in this area — but in being able to get information which you had no reason to expect to want.

There are other reasons why I am very wary of the traditional specify-design-code-debug procedure.  For one thing, it is costly and for another there is usually a vast sea of miscomprehension which separates the potential user and the DP professional.  In practice the whole procedure is a reiterative cycle, and a lot of time is spent on detail some of which turns out to be irrelevant.  I prefer implementation by building prototypes inhouse which gives you a chance to try out your design before spending too much time or money and to refine as the mood takes you.

Apart from systems like Topic, there are numerous tools included without extra cost in a standard development platform implementation of UNIX which help you to structure, manipulate and retrieve text based information.  People often make very silly criticisms of UNIX:  one is its sheer size.  In fact the inclusion of numerous software tools, all designed to work happily with one another, is one of its greatest merits.  On other systems many of those good things must be acquired as expensive add-ons.

One other relevant point is that the size of administrative data bases for most Solicitors (apart from accounts) is relatively small, and I am not at all sure that you really need the facilities of a powerful relational data base manager optimised as they are for fast Boolean operations, projections, joins and the like; simple versions of which are in any case included in UNIX.

So we do not have a relational data base engine in our firm, but I suppose that since fashion dictates so many things we will get one one day.


## 9.11   SECURITY

Machine failure, powercuts and viruses are the great Enemy and there is very little you can do about them except to make as many backups of your data as frequently as your time and money will allow.  In that way, though you cannot prevent, you can limit the amount of damage which they cause.

This is a far more effective guard against viruses than pandering to the natural paranoia of system administrators by unreasonably restricting access to machines and their use.

On the question of confidentiality all password and similar schemes fail at some time or another for reasons which have less to do with the machine or the security scheme than with human behaviour.  I would highly recommend the paper published by Bell Laboratories on the subject of passwords.

In my view the only way to keep really confidential information confidential is either not to put it onto the global system at all (e.g. keep it on locked away floppies) or to encrypt it, an easy enough thing to do on the computer as long as you don't forget the secret key! All codes appear to be breakable given enough time and effort so you need to re-encrypt the data from time to time.


## 9.12   ARCHIVING

The most common current approach is to work with a fixed amount of secondary memory — hard disk for example and as this fills up either to erase those items which are considered not to be worth keeping or to move them onto cheaper off line storage media such as tape or to some combination of these. There are several difficulties in this approach.

It is in the nature of our profession to be uncomfortable about destroying information and whilst various firms adopt different policies about the time for which papers should be kept, it is my experience that what actually happens in each firm differs widely from one individual to another. The difficulty of deciding is aggravated for computer data because the invidious decision has frequently to be made considerably sooner than the decision about the file itself. Then there is the fact that it is sometimes the file that you thought was eminently destroyable which turns out to be the one file you wished you had kept. As I said earlier a chief aim of the system should be to be able to find information you had no reason when you acquired it to expect to want when later you do want it. One mega firm that I once worked for solved the problem by insisting that a "bible" (i.e. a separate bound copy of all the salient papers) was made for each transaction before the file was put away, and kept back when the file went away. This was a time consuming business but it was worth it: they were almost invariably large transactions. Clearly this would be wholly unsuitable for debt collecting and running down cases.

We have found the DAT and Cartidge tapes we use to be somewhat unreliable. This is not so bad when you are making the backup: you can always put in another tape. But no amount of verifying that you made a perfect backup will ensure that the tape reads back correctly when you need to restore. So you cannot afford to take chances. You have to back up very frequently.

Tapes are a sequential medium: files are stored in one long line, and you cannot read any file without first reading each file which stands before it in the line. So it can take a long time to restore a file at the middle or end of the tape. You might be tempted therefore to back up onto shorter rather than longer tapes. You will then end up with shelves of backup tapes, with catalogues and all the problems of finding the tape which has the file your user needs.

It's not a good idea to make your backups while the system is in general use. While the backup is running your users are creating new files or changing existing ones, and it then becomes difficult to restore the status quo at any particular time. A decent operating system like UNIX will allow you to run your backups automatically late at night or in the very early morning, but you need a human being to leave a tape in the machine and take it out after the backup is done. So you will in fact tend to want to use longer rather than shorter tapes, unless you are prepared to spend money on fitting your machine with mutiple tape drives and organise your file system into distinct separate physical parts each of which will fit onto one tape without overflowing.

In the result you end up with a posse of administrators and your users have to wait before they can access the files they need.

This is not our approach. A modular system like ours lets you go on adding devices and machines virtually *ad lib.* So we can think of the system as having infinite secondary storage. But this doesn't allow us to forget about archiving altogether. For one thing, it is pointless making security back ups of files which have not, and will never, change. We believe that archived files should be kept on on-line random access storage at least for as long as we would have kept the corresponding paper files. This obviates the delays inherent in using tape as on a random access medium you can for all intents and purposes find any file as quickly as any other. The simplest method is to use hard discs. This is I believe much cheaper than tape. We can buy a 1.3 gigabyte drive for about £2000. So you can buy eight or so gigabytes a year for the cost of employing one administrator whose job it is to backup and restore files on tape. Of course, the figures need adjusting to take account of the fact that you can't go on adding disk drives to a single PC *ad lib* — but they point in the right direction.

It is only too easy by mistake (or for that matter by malice) to delete archived files. So we plan to use WORM drives rather than Winchesters. These have the advantage that when and if we decide to take an archive off line we can simply remove the disc. For the moment we have only one WORM drive and are investigating how best to have everything online. Jukeboxes seem quite expensive and it might well be cheaper to daisy chain a number of WORM drives off a SCSI controller.

One problem with WORM drives on UNIX, is that every time it accesses a file it updates the file access time which is stored with the file and the disc can run out of space even though no new file has been added or any file changed. The solutions we have come across so far are too costly. We have not yet found a solution at an acceptable price,

## 9.13    CONCLUSION

I realise that some of my views may be considered eccentric, but I hope that you have found some interest in them.  If not, can I at least wish you as much fun and satisfaction out of your system as I have had from ours. Thank You.

# 10. Design and Implementation of Litigation Support Centres in the Law Office

*Patricia S. Eyres*
*Litigation Management & Training Services, Inc.*

## 10.1  THE GENESIS OF AUTOMATED LITIGATION SUPPORT

Computerised systems for litigation management began in the United States with very large antitrust cases in the mid 1970's. The prohibitive cost of the computer hardware, coupled with the administrative challenges inherent in managing a massive document coding operation, made it impractical for law firms to direct the litigation support effort internally. Thus, most projects were performed by experienced service bureaus. The result was the birth of a diverse and responsive industry, with vendors of automated litigation support services acquiring the expertise to manage massive information databases for the legal community.

As the advantages of computers became more apparent, litigation automation projects were soon the source of extensive Requests for Proposal. The successful bidder ordinarily handled the entire litigation support effort; from early collection and organisation of the document repository, through database design, document coding, quality control, data entry and database maintenance. Coordination of project details with counsel were limited to initial database design and preliminary controlled vocabulary development. Vendors also offered search and retrieval services, although the databases were available to litigators over the telephone lines using a modem.

For many years, litigation support was managed most cost-effectively by external service businesses. The cost of computer hardware alone was prohibitive for most law offices. Even firms with mainframe systems were reluctant to devote the resources necessary to housing sizeable litigation support databases. Corporate counsel were also concerned about breaches of security and waivers of privileges associated with legal databases housed on company computers.

*V. Mital (ed.), Advanced Litigation Support & Document Imaging, 93–125.*

Outside counsel were equally unwilling to dedicate the necessary time and resources to the document coding effort. So too, law offices were ill-equipped to manage very large document repositories. Lacking the space to manage the documents and the personnel to process the information, firms elected to refer the projects to outside companies. Litigators relinquished control over their documents. The costs were customarily passed directly to the client, with the law firm foregoing any profit beyond the fees they generated for communications with the litigation support vendor and their own searches of the database.

Times have changed. Personal computers have paved the way for law offices of every size and practice area to make productive use of litigation support techniques. The 1980's ushered in sweeping technological advances which have transformed computer-assisted litigation support into a viable in-house option for most law offices. The next decade promises to rival current technology with sleek, powerful and affordable systems which will integrate every aspect of information management, enabling trial lawyers to control their cases entirely in the privacy of their offices.

Many cases are still too large for effective in-house management of the entire automation project; thus, litigation support vendors still provide a valuable service. However, law firms and corporate legal departments now make productive use of internal Automated Litigation Support Centres ("ALSC") for many of the cases they regularly handle. Creative firms have begun to manage most cases internally, while still using outside services to supplement their own computerised capabilities when preparation time is severely compressed or the document population is too large. This, coupled with internal computerisation of the firm's work product retrieval system is a valuable — and potentially profitable — adjunct to the firm's litigation practice.

## 10.2   BENEFITS OF IN-HOUSE LITIGATION SUPPORT CENTRES

Successful law firms and corporate law departments recognise the unique benefits of instituting ALSCs internally in the law office. These separate units, trained to perform specialised litigation support procedures, enhance the firm's ability to master its information management objectives and to control litigation support expenses.

The size and structure of the firm's litigation support mechanism depends largely on the volume and complexity of the automation projects it will administer. Whether the litigation support function is disbursed throughout the firm, monitored by individual attorneys or unified into a distinct division serving all litigators, the ability to serve clients through internal computerisation of cases is critical to the competitive posture of law practice in the '90s and beyond.

An ALSC in the law office serves the dual purpose of increasing economies of scale and facilitating an additional litigation profit centre. It also gives the lawyers an aggressive advantage over their adversaries in pending litigation and the firm a competitive edge in client development.

## 10.2.1  Profit Centre Capabilities

Litigators are turning more frequently to litigation support capabilities as an additional profit centre. In addition to the customary fees for professional services, which may include attorney and paralegal work on the litigation support databases, firms charge their clients for the document management, coding, full-text enhancements and related litigation support services accomplished by their non-lawyer ALSC personnel. Just as the time share companies and litigation support vendors are handsomely compensated for their substantial efforts, law firms may directly invoice their clients for the computer-assisted techniques they implement. This permits the firm to provide highly competitive legal services, while at the same time diversifying its sources of profit. Indeed, the client benefits from increased cost controls, enhanced security and better command over its documents and evidence.

As an alternative approach to billing for in-house litigation support services at a profitable rate, many firms choose to develop a fee structure designed to recoup only the expenses of the litigation support effort. This may include reproduction services (photocopying and microfilming), adjunct services (court reporter deposition diskettes, scanning, optical character recognition, image capture or keying services for full-text databases), as well as firm overhead attributable to litigation support, such as computer leases, document repository management (storage fees) and overtime expenses for data entry personnel. These fees are passed to the client directly as a disbursement. When the services take place in-house, such as photocopying, scanning, word processing or computer technical support, the fees are set at sufficient levels to meet direct expenses only. This saves the client money and still permits the firm to provide "value added" services in the form of automated litigation support. Legal services for document analysis, on-line deposition enhancements and related work are billed in the customary manner.

## 10.2.2  Adjunct Client Services

Still another option available with the implementation of an in-house ALSC is the ability to offer imaginative services to institutional clients. In the United States, law firms that represent manufacturers of products involved in nationwide litigation use their internal computer systems to manage large document productions and repositories and testimony databanks for their clients. Similarly, employers with repetitive personnel lawsuits use litigation support systems quite productively. This places the litigators in a position to provide extremely cost-effective legal services. With greater familiarity over the subject matter and knowledge about the documentary evidence relevant to their client's litigation, they increase their value as a resource to the corporate counsel who retain them.

### 10.2.3  Practice Development Opportunities

Internal litigation support capabilities are also becoming a reliable resource for practice development. The law firm that demonstrates to potential clients the power and skills to manage their litigation — and to do it cost-effectively — is well positioned to capture a competitive share of quality business. Likewise, the in-house law department which is better able to manage its own documents and resources is able to limit its reliance on outside counsel for document and discovery management and reduce expenses. Accordingly, both public and private law offices are turning to in-house litigation support centres as a means to achieve control and to increase profitability.

### 10.2.4  Paralegal Career Advancement

Large, complex litigation is labor intensive. Structured litigation support units within the law office provide a valuable opportunity for firms to offer a rewarding career path for skilled legal assistants and other employees who consistently perform the significant "back-up" functions of discovery and trial preparation.

In-house litigation support centres provide precisely the opening for the enterprising law firm to offer growth potential to their valued para-professionals. Legal assistants who serve the firm in a litigation support capacity may look forward to developing management skills and other challenging career opportunities, while still using their legal talent. A carefully administered ALSC provides the automated legal assistant with the potential for promotional opportunities, starting with supervision of the coding staff, moving up to leading an automation project, then managing an entire database, and finally to administering the firm's entire litigation support unit. **Figure 1** reflects a sample structure for one litigation support centre from a mid-sized law office.

## 10.3  PLANNING FOR IN-HOUSE LITIGATION SUPPORT

Litigation is rarely predictable. The twists and turns of pre-trial discovery can adversely impact even the most liberal budget. Litigation support costs and document management can quickly rage out of control. Consequently, in addition to comprehensive management of the databases themselves, appropriate administration of the responsible litigation support unit is pivotal if the ALSC is to be a profit centre.

In order to realise the substantial benefits of an ALSC, careful planning, detailed organisation, appropriate staffing and efficient administration are vital. This paper will focus on each of these elements and provide a checklist for firms to use in designing an in-house ALSC.


## 10.4   FOUR-STEP PROCESS FOR IMPLEMENTATION OF AN ALSC

### 10.4.1 Step 1:   Design the appropriate structure for the ALSC

Most litigators who are exposed to automated systems supporting their litigation find the techniques so useful that they are anxious to implement them on more than one case at a time.   In this way, the costs associated with hardware and software acquisition may be spread over many cases; thereby increasing the economies of scale and reducing the costs per project for each client.   Indeed, once the firm invests in hardware and software systems, it should explore appropriate ways to utilise those resources effectively and at the same time to recoup the capital investment.

Firms planning to manage multiple cases generally adopt a formal structure, as depicted in **Figure 2**.  While the number of staff members will vary depending upon the work flow and substantive complexity of the projects, a basic organisational framework is necessary.

### *10.4.1.1 Single Project*

If the impetus for the litigation support database is a single large case, standard document management techniques and text retrieval systems are appropriate. Existing support staff may be pressed into action and a conference room or spare office donated to the project.   The time expended setting up suitable computer systems and training the firm's personnel will be absorbed by the firm or reimbursed directly by the client.  Hardware and software may be rented and interim procedures developed for handling the special project; sometimes with the assistance of a consultant.  Expenses which are directly related to the project may be passed on to the client.  When the project is completed or the case ends, the firm may return to its manual systems.   Thus, a formal ALSC structure, with the concomitant procedures, personnel management, profit centre policies and administrative control, is unnecessary.

On the other hand, if the documents are voluminous, if the issues are intricate or if the time frame within which to implement the system is restricted, the trial team will need to adopt many of the same methods inherent in a structured ALSC.

*Staffing*

Personnel are the first consideration. The firm should employ additional staff in a long term temporary capacity. Litigation support, to be effective, is extremely detail oriented. The procedures for building quality databases require undivided attention. It is especially important when time is short to minimise reliance on existing firm support staff (paralegals, word processors etc.) because their ongoing responsibilities on other cases may distract them from a concentrated approach to the automation project.

In order to effectively manage a single large project, the staffing considerations are similar to those utilised in structured litigation support centres. Therefore, the reporting structure and staffing options discussed in this paper will be quite useful. **Figure 3** is a sample organisational structure for a litigation support project staff.

It is essential to designate an overall manager of the project, who will be supported by coders and data entry personnel, as well as middle level supervisors who perform quality control and monitor the progress of database design and implementation. The project manager may be a trusted paralegal; however, the individual's time investment will be extensive so hourly billing rates should be commensurate with these responsibilities. The project can quickly become too expensive when a very senior legal assistant is assigned the role of Project Manager. Likewise, the project manager should be able to devote complete attention to the case, without conflicting time pressures.

Law firms in major metropolitan areas have utilised computer-assisted litigation support systems for several years, and have had the opportunity to develop experienced and skilled staff. Many legal assistants and experienced litigation support personnel, who were exposed to automated systems in large law firms and corporate legal departments now provide freelance services as project leaders. This may be an exciting opportunity for the automated paralegal, but it also provides a resource for the firm seeking qualified temporary assistance with a single large case.

When the firm lacks internal experience with computerised litigation databases, it is prudent to recruit a project manager with experience on similar automation projects. This individual will work directly with the trial team and coordinate with the designated senior paralegal or attorney, thereby facilitating the processes of database design, vocabulary control and database management.

The costs of recruiting personnel, administering temporary payroll, training existing staff and developing litigation support personnel procedures for a solitary project — as opposed to a stable workforce handling multiple case databases — may be significant. The justification for the expense will depend upon the size of the task, the information management objectives, the ability to handle the case without computerisation, and the client's budgetary constraints.

*Facilities*
With a very large project, the trial team must also consider appropriate facilities. Most law firms pride themselves on their attractive, professional surroundings. This is typically expensive space. The physical requirements of managing a large-scale computerised litigation support project may be incompatible with the firm's existing offices.

Automated litigation support involves physical management of crates, cartons and cabinets loaded with paper. Secure control over documents distributed during coding and quality control is vital. Data entry frequently occurs on multiple workstations. The trial team cannot tolerate lost or disorganised documents or breaches of security. The litigation support effort will be dysfunctional if the coding and data entry space is inadequate. Thus, when affordable temporary space in the firm's existing office building is unavailable, the coding may have to be moved to a convenient area close to the firm's office.

*Lawyer Commitment*
Experienced litigators who are exposed to automation for the first time may be uncomfortable with the different tools and document control procedures. They may be sorely tempted to return to the security of familiar routines. When the inevitable time pressures arise, they may wish to avoid spending time to help design the database and develop vocabulary controls. They may also try to bypass the opportunity to learn how to use the system most effectively. When this occurs, the litigation support effort is doomed.

Once the decision to automate has been reached — either because it is desirable or unavoidable — the litigators must be prepared to devote attention to the planning process and to administer the project. Attorney involvement is absolutely essential on both substantive and procedural levels. The responsibility cannot be delegated to the paralegals alone; regardless of their experience and competence. The trial attorneys must devote time to meet with the litigation support staff to assist with database design, to provide invaluable input into vocabulary controls and useful reports/printouts. If they fail to do so, they cannot expect a database responsive to their information management needs.

Litigators who lack experience with automated systems need not become instant experts to make productive use of in-house litigation support on a specific case. Certainly, by recruiting experienced litigation support professionals who can direct the discussions on database design objectives and controlled vocabulary development, the lawyers vastly reduce unproductive "wheel-spinning." While the Litigation support manager is trained to facilitate decision making, the lawyers **must be prepared** to spend the requisite time focusing on their information management objectives and database reporting requirements.

In addition to pledging time at the outset of the project, the litigators should expect to devote some effort learning to use the system. Training time for members of the trial team is frequently limited to search and retrieval and very modest data entry in attorney notes fields in document management databases and annotations to full-text materials. Training may take place outside the normal hectic work week. Through directed training, the litigators may focus on specific search and retrieval techniques which will allow them to master the information, to access it wisely and to pinpoint vital information instantly. Placing control of significant case information at the lawyers' fingertips is a powerful method for ensuring the ultimate success of the endeavor.

### 10.4.1.2  Internal Systems in Mid-Sized and Large Firms

ALSC structures in law firms are as diverse as the substantive areas of many litigation practices. When more than one case or attorney will be using the litigation support systems, some level of structure is advisable.

More formal approaches frequently provide significant advantages; particularly when the structure reduces costs, promotes stability in trained personnel, increases uniformity and enhances profitability. With the availability of cost-effective litigation support systems for use throughout the law office, many firms have found it productive and economical to establish systematised approaches to their in-house litigation support efforts.

As noted earlier, a law firm need not designate a separate "unit" as an ALSC. However, if the firm wishes to develop litigation support as a profit centre or to realise the economies of scale which exist when a stable group of trained personnel administer all the firm's automated litigation support projects, some elements of an ALSC may be appropriate.

The level of bureaucracy will depend on three major factors:

- how often the ALSC will be used and for how many separate cases or projects
- the size and complexity of the cases
- whether the firm intends to use the ALSC as an additional litigation profit centre

The starting point is a system manager who is responsible for overall administration of the ALSC. This individual is frequently given the title Litigation Support Manager. It is strongly recommended that one individual be given the responsibility to manage the litigation support centre, while reporting directly to a managing partner or firm committee. This enables the firm to develop consistent policies and to assure that the manager is provided the authority to implement and enforce the procedures.

The determination of when and whether to use multiple project managers, coding supervisors and other mid-level supervisory personnel depends exclusively on the anticipated work flow the ALSC will enjoy.    If the firm has a large and substantively varied practice, and expects to automate all or the majority of its cases, it may require several project managers and related staff to meet the intense demands of building and administering the databases.

For example, a firm which handles large construction, environmental, professional liability, intellectual property or real estate litigation, may productively use multiple project managers, with each seasoned in a particular substantive area. This will maximise the effectiveness of repeated database design and coding convention/vocabulary control activities.

The larger the cases or the more databases the ALSC will handle concurrently, the greater is the need for experienced, stable and competent permanent staff. Thus, the firm may choose to employ a core group of Project Managers on a permanent basis, along with the litigation support manager.

Coding and data entry staff, on the other hand, are frequently hired on a temporary project basis; both long term and short term.  The flow of work fluctuates most often with document coding, as receipt of documents may be delayed by unforseen circumstances and/or disposition of discovery motions.   When the documents do arrive, the time for processing through the ALSC may be severely limited, requiring the litigation support manager to recruit, train and supervise additional temporary coding staff.  This is also true for quality control and data entry personnel.  The ability to locate competent, reliable temporary litigation support staff — sometimes on very short notice — and to integrate them into the efficient operation of the ALSC requires skill and experience.  However, when big cases suddenly settle and coding immediately ceases, it is beneficial to be able to cut off expenses at once by discharging the temporary personnel.

One drawback to using temporary or project staff consistently is that the costs of recruitment and training are incrementally higher.  Firms which have a steady flow of automation projects or which can integrate their coding and quality control staff into other ongoing automation projects (including work product retrieval) find it advantageous to employ a small group of steady reliable coders/quality control specialists as permanent employees or long-term temporary workers.   When additional work requires up-staffing, more coders can be hired on a short term project basis.

### 10.4.2  Step 2:  Select staff and determine their responsibilities

The litigation support manager is the most important staff member.  He or she should have experience with automated systems; preferably with the firm's database software. However, an experienced individual can be trained on different software. It is much more critical for the manager to have solid organisational skills and detail orientation.

In addition, the ability to interface well with both subordinate staff and the trial team is extremely useful. Finally, good presentation skills, including the ability to effectively describe ALSC capabilities and inspire confidence among clients and litigators, is a good skill for the successful litigation support manager.

The manager should be an experienced paralegal, since database design and development of vocabulary controls is best accomplished by someone with both legal and litigation support experience. This is especially valuable when automation is new to the lawyers.

The role of the litigation support manager is varied and critical. **Figure 4** is a sample job description for a litigation support manager. In addition to the customary administrative responsibilities of managing personnel and work flow, the litigation support manager serves as the primary liaison between the ALSC staff and each litigator or trial team. The manager often serves as the principal spokesperson in promoting the use of automated litigation support throughout the firm. Thus, the best resource for internal promotion is a skilled paralegal or administrator who has earned the confidence of the litigators.

Firms using project managers find that the same skills and qualifications apply to the selection process. The checklist accompanying this paper contains a comprehensive list of qualifications for selection of both litigation support managers and project managers.

*Database Design*
Because the manager is directly responsible for insuring that the firm's litigation support procedures are consistently developed and uniformly enforced, the responsibility for coordinating with each trial team in the initial design of the their database structure, coding conventions database output objectives and report formats rests with this individual. In large firms or when the workload is too burdensome for the manager alone, individual project managers may successfully fulfill this function, reporting directly to the ALSC manager. This facilitates consistency, and at the same time permits ALSC personnel to work with different litigators. The benefits of wider experience on different substantive cases benefits both the trial team and the ALSC personnel.

*Specialised Subject Matter Project Managers*
Some project managers specialise in specific areas, such as work product retrieval, construction litigation, commercial disputes (such as bankruptcy or savings and loan litigation), product liability or professional malpractice. This assists the firm and the

ALSC staff in concentrating more cost-effectively on development of database design and vocabulary controls that are targeted to the type of litigation. The wider the project manager's experience in a particular area, coupled with extensive exposure to automated systems design, the more productive will be the development of new databases. **Figure 5** is a position description for a project manager. Project managers report to the litigation support manager or, in smaller structures, directly to the trial team.

*Budgeting*
Budgeting and forecasting is an essential component of the litigation support manager's position. When a new case or project is assigned to the ALSC, the manager is in the best position to develop a preliminary budget. The budget takes into account how much time the project will consume, how the client will be charged for the services, and the total anticipated litigation support expenses. In turn, the manager works directly with the trial team to assure adherence to the budget or to flag instances when the nature of the work will exceed the agreed upon budget. project managers should also be able to perform this function.

For firms using the profit centre approach, the manager also oversees the ALSC timekeeping and document abstracting fees. The manager will approve the bills, subject to final approval by the billing attorney.

*Personnel Management*
Supervision and performance evaluation of ALSC personnel is another function of the litigation support manager. This is a management function which generally is not performed by paralegals, so finding someone with the requisite skills and experience may be a challenge.

*Coding And Quality Control*
The principal role of the ALSC is to provide coding services for document management databases and to provide enhancement services for full-text retrieval systems. Thus, the role of the document coder is among the most important in the ALSC. **Figure 6** is a representative position description for a document coder, sometimes called a "document analyst." Document coders are responsible for reviewing the documents, completing the appropriate coding forms (or completing direct input into the computer if on-line coding is the chosen method for the ALSC or a specific case). If on-line coding is used, the coders must be trained not only on document coding techniques, but on the data entry and validation provisions of the software. They must also have basic typing skills, although speed is rarely a critical function.

The best document coders are individuals with good vocabulary and a college education, but no direct legal experience. This is because objective attention to detail is critical. Document coders are instructed to record objective material from the text of the documents, with as little subjective analysis as possible. This assures objectivity and consistency. When law students or experienced paralegals are used as document coders, their training in legal analysis frequently interferes with their ability to objectively record bibliographic data from documents.

In addition, document coding is extremely tedious. It requires excellent language skills and unwavering attention to detail. For this reason, successful firms limit the work shifts of document coders to four or six hour increments.

The document coders report to their project manager, from whom they receive their daily assignments. They do not report to quality control specialists, although the latter will directly review their work for errors. The Project Manager is responsible for advising document coders when their work requires improvement, and for monitoring the rate of coding errors from individual coders and quality control specialists.

### 10.4.3   Step 3:   Design Systematic Procedures for the ALSC

Once the organisational structure of the ALSC is established, staff is selected and responsibilities are assigned, the firm should establish policies and procedures for administering the ALSC. These procedures should be well planned and capable of consistent enforcement.

*Permanent Personnel*
Of course, the firm's standard personnel policies, compensation systems and other administrative procedures will be applicable to the permanent staff hired for the ALSC. Indeed, if the firm chooses to do so, all personnel administration — except performance evaluations — for ALSC staff may be handled by the firm's administrator, rather than the litigation support manager. The firm should attempt to fully integrate the permanent employees involved in litigation support with the remainder of the firm; especially if they are newly hired.

*Temporary Staff*
Most law firms do not utilise temporary personnel in professional or para-professional capacities. Consequently, firm management may have to address the appropriate method to compensate and administer payroll for the temporary coding and data entry staff. This is particularly an issue when work schedules will vary week by week.

*Authority and Communication*
Above all, distinct lines of authority are required, so that redundant work is limited and quality control is maintained. The litigation support manager will have overall authority to control the flow of work, the security of the facility and the consistent

work product of the ALSC.

Clear lines of communication are also essential. It is certainly desirable for the ALSC to be kept busy, with ongoing projects and multiple databases/document repositories to manage. When this occurs, however, effective organisation and communication become even more important. Each trial team using the services of the ALSC must feel confident that their databases are being administered in a timely, conscientious, quality controlled and secure manner. On occasion, imminent time deadlines, coupled with the competing demands of multiple litigators will sorely challenge even the most organised ALSC manager. Effective communication is necessary to assure that the ALSC runs smoothly.

*Document Processing*
Physical management of the facilities merits some attention. With multiple document repositories, space is always at a premium. However, nothing is more disruptive to the process than for lawyers or ALSC staff being unable to locate the physical documents following their identification in a database search. Likewise, the physical flow of the documents through the coding and quality control process is absolutely critical. Thousands of documents will pass through the ALSC every month. Many will be dropped on the doorstep following a document production or deposition. The documents must be segregated, numbered, segregated into manageable batches and distributed for coding. They must then be segregated for quality control, data entry (if not coded on-line) and placement in the document repository. The process is more complex when privileged or otherwise confidential documents must be segregated for coding and then returned to a different location than the open repository.

The process, even for a single case, demands precision and security. Multiply the process by five or more ongoing projects and imagine the chaotic consequences without extensive procedural controls. **Figure 7** represents a diagram of an organisational chart for processing documents in an ALSC; **Figure 8** is a sample batch log for maintaining up-to-the-minute information on the status of every document under the control of the ALSC at any point in time.

In addition to management of document processing among multiple projects if the ALSC is to operate as a profit centre, the firm should address methods for monitoring time and other services to be charged to clients. This is ordinarily administered by the litigation support manager.

*Search And Retrieval*
Search and retrieval of database information is the heart of the litigation support process. The ALSC must make this easy for the lawyers to accomplish. Immediacy is also a factor. Thus, well planned methods of database and document repository access should be in place. Attorneys wishing to search and/or retrieve a hard copy

should be trained in advance on how to accomplish these tasks; but ALSC staff should also be prepared to offer immediate assistance.

Attorneys from remote locations, such as branch offices, must receive appropriate training and be assigned passwords for access to their databases. The litigation support manager and project managers are then responsible for monitoring the frequency and results of access on the phone lines.

*Computer Maintenance*
A busy ALSC may operate as much as 18 to 20 hours each day, including weekends. Thus, it is essential that the hardware and software work effectively. Large firms often have the luxury of full-time technical support staff. Many firms do not. Thus, the manager must maintain a relationship with a reliable consultant or vendor and current service contracts on all equipment.

*Back-Up Procedures*
The final aspect of ALSC procedures should be sound data back-up procedures. Data input, editing, enhancements, annotations, search queries and reports that litigators have saved on disk all may be wiped out in an instant. The lost data may represent days, perhaps months, of concentrated work. The client may already have paid for the services. Disaster.

The one policy an ALSC must enforce rigidly is data back-up. In large departments, this should be done several times a day. Daily and weekly back-ups should be kept both on-site and off-site. The manager must take personal responsibility for this critical component.

As firms in San Francisco, California, discovered following a large earthquake, a disaster (fire, flood, earthquake) may temporarily prevent the ALSC from operating for days or even weeks at a time. Back-up policies should include a contingency plan for access to another firm or client's computers for search/retrieval if the firm's computers are not available. This requires that a master set of the database software be stored off-site in addition to up-to-date data back-ups.

### 10.4.4  Step 4:  Determine whether the ALSC will be a Profit Centre

Litigation support activities are time consuming. Personnel expenses are significant. Computer equipment is a major capital expenditure and rental of additional facilities is costly. Thus, the law firm must price automated litigation support services at a rate sufficient to assure that the ALSC does not become an unacceptable drain on firm resources. A self-supporting litigation support centre is crucial. A profitable litigation support centre is not essential, but can be a valuable resource for expanding litigation practice in the '90s.

One of the more troublesome financial aspects of managing an in-house litigation support centre is obtaining consensus among the partners concerning policies and procedures for billing ALSC services to clients. Some litigators have a natural reluctance to add any additional items to their clients' already spiralling litigation expenses, particularly when the description of services will clearly be beyond the traditional hourly fees for professional services. Others are not convinced that computer-assisted litigation support is cost-effective. Still others are concerned about appropriate controls over computerisation services. For these reasons, it is often **essential** for the firm - through the litigation support manager -- to make informative presentations concerning the capabilities of the ALSC and the methods in place to assure control over expenses. Likewise, the firm can use this approach to make similar demonstrations to clients. When presented professionally, by explaining the utility of information management and highlighting the reduction in duplication of expenses, this serves both an educational and promotional purpose.

Once the decision is made to charge clients for specific ALSC services, either to recoup costs or to produce additional profit, the firm should establish a consistent policy concerning the services which will be billed, how fees will be determined, and who will be responsible for monitoring the ongoing expenses. The litigation support manager and/or project manager will be a valuable resource in setting a preliminary budget and comparing actual expenses to the budget on an ongoing basis.

Most firms delegate the task of reviewing the time sheets and bills to the litigation support manager, who interfaces directly with the billing attorney.

*Database Design/Planning/Implementation Services*
The determination of hourly billing rates and billing policies is important. The services of the LItigation support manager and/or project manager on database design, development of vocabulary controls and case-specific coding conventions, preparation of coding manuals and user guides, and development of report formats are customarily billed as hourly fees. The description on the billings is case analysis, system design, database configuration and may include time for document review and analysis.

*Full-text Database Services*
Similarly, the time involved in constructing and maintaining full-text databases; specifically, design and input of vocabulary controls, on-line summaries, search/retrieval and report formatting are billable in the traditional hourly fee structure. Naturally, search and retrieval, development of sorted reports and related services performed by attorneys, paralegals and law clerks are billed at regular hourly rates.

*Document Management Services*
Document databases are frequently treated differently because it is **extremely**

difficult to track and monitor the time expended on physical document control and numbering, document batching, coding, quality control and data entry. Coding services include the following:

- receipt of the documents and placement of an internal document number, segregation of originals.
- coding of each document for information required by the data fields
- ongoing quality control, supervision and review of the coding effort to ensure consistency
- specific quality control procedures to ensure accuracy, including spell checks, name, date and vocabulary validation
- batch loading of the database, daily backups, weekly back-ups and monthly back-ups.

The very value of using an ALSC for document management databases is that the type of work can be performed at a much lower cost by personnel specifically trained to analyse, code and input the information. This results in dual benefits: the litigator's time is freed to perform case analysis and select strategies and the significant costs of document management for the client are controlled. Thus, charging these services by the document is the preferable method. Thorough planning, database design and implementation are critical to success. When this is accomplished, construction of document management databases can be profitable.

Accordingly, most firms bill litigation support services involving document management databases by the document or the page, patterning their billing practices and rates after those utilised by external litigation support vendors. Exceptions in favor of hourly billing for document coding arise when the subjective issue coding is so complex that it must be accomplished by lawyers, law clerks or experienced paralegals or in small firms where all database management activities are performed by the legal professionals. These latter situations dramatically increase the costs for document management databases, but may be justified under the unique circumstances.

Once the firm decides to bill document management services in this fashion, it must establish a fee structure. Per document fees vary extensively, depending upon these factors:

- the number of documents
- the level of coding required (bibliographic only, bibliographic/objective or subjective)
- the complexity of the subjective coding
- the time frame within which to design and implement the database.

The average document is 3.4 pages. Obviously, letters and other correspondence will be shorter and agreements, transactional documents and related material much longer. However, the shorter documents are generally more coding-intensive, since they contain numerous references to names, dates, documents and events. When setting fees, the litigation support manager will determine an "average" per document charge, taking into account the anticipated length of time it will take to process the average document. This analysis includes not only physical receipt, development of coding "batches," document coding and analysis, quality control, data entry and necessary editing, but also all procedures for maintaining the physical document repository.

The complexity of subjective coding is another meaningful factor when setting coding fees, since the more detailed issue/subject coding requires more intense analysis of each separate document and greater quality control. When experienced coding personnel are used to perform bibliographic/objective coding, the number of documents which are reviewed for quality control are significantly less than when complex subject coding and extensive key words are used. Increased scrutiny during the quality control process involves additional time and expense of quality control specialists. Fees are set accordingly, again with substantial consideration given to the anticipated level of coding and necessary quality control. **Figure 9** contains a sample letter to a client regarding litigation support fees to be generated by an ALSC.

Establishing a profit centre for litigation support is a challenging process. The firm's management must reach agreement on policies and procedures, and the process must be monitored extensively to assure effective cost controls and appropriate billings to clients. **Figure 10** is a sample memorandum regarding a firm's ALSC billing policies and procedures. The effectiveness of the policies and their consistent enforcement should be assessed regularly during the first year the firm bills to clients all or any part of services performed by an ALSC. There are several methods to determine whether the use of an ALSC profit centre is appropriate and successful for the firm:

- the success results of the ALSC
- the reactions of the firm's clients
- the relationship between preliminary ALSC budgetary determinations and actual expenses on the project
- the costs of administration
- forecasts of the long term profitability of the ALSC
- the actual profitability of the ALSC.

## 10.5   CONCLUSION

As competition for quality business increases, as litigation grows more complex and as courts continue to reduce the time within which lawsuits must be resolved or tried, the use of computers will become even more necessary to the successful litigation practice. Automated litigation support centres in the law office are a practical and realistic method to meet the challenges the coming decade will bring and to profitably achieve information control.

The following checklist is designed to facilitate a step-by-step approach to the establishment of an ALSC in the law office.

### PLANNING CHECKLIST FOR DESIGN AND IMPLEMENTATION OF LITIGATION SUPPORT CENTRES

In order to realise the substantial benefits of an ALSC, careful planning, detailed organisation, appropriate staffing and efficient administration are vital. This paper has focused on each of these factors. This checklist is designed to aid the process of establishing an ALSC.

### WILL THE LAW OFFICE BENEFIT FROM AN ALSC?

1.  Does the firm regularly handle document-intensive litigation?

    a.  Are the cases distributed among multiple attorneys and legal assistants?
    b.  Are the files and documents centrally accessible to all who need them?
    c.  Are attorneys who share case responsibility able to effectively search and retrieve information?

    (1)   Do the attorneys and legal assistants use consistent methods for indexing and storing documents?
    (2)   Do the practice areas include repetitive substantive issues?

    d.  Are the documents and document indices housed centrally or throughout the firm?
    e.  Does the firm have a consistent method for billing document indexing time to the client?
    f.  Does the firm have a consistent method for billing in-house document organisation and management services to the client?

2. Does the firm regularly handle testimony-intensive litigation?

   a. Are all or a majority of the deposition transcripts regularly summarised?
   b. Are all or a majority of the trial transcripts regularly summarised?
   c. Does the firm regularly handle appellate matters involving transcripts on appeal?
   d. Are all or a majority of the appellate transcripts summarised? Indexed? Catalogued?
   e. Do the firm's attorneys and/or legal assistants regularly share summaries of testimony from multiple cases in the office?

      (1) Are the summaries of testimony prepared with consistent formats?
      (2) Are the summaries of testimony prepared with consistent terminology?
      (3) Are summaries of depositions and other testimony searched for keywords and phrases? Are summaries prepared with this in mind?

   f. Does the firm maintain a master index of transcripts from recurrent expert witnesses and clients?

      (1) Is the index updated regularly? Does it include open cases only, or both open and closed cases?
      (2) Is the index readily accessible to all interested legal professionals in the firm?
      (3) Is the index sufficiently complete to permit the user to quickly determine whether testimony from a particular individual is available within the firm?
      (4) Are the transcripts themselves readily accessible?

   g. Does the firm have a consistent method for billing in-house services for deposition digesting and transcript summarising to the client?

3. Does the firm handle multiple matters for institutional clients?

   a. Does the firm maintain a master index of documents produced by the institutional client in multiple cases?
   b. Is the firm able to provide the client with an index of documents and/or transcripts produced in multiple cases?
   c. Can the firm provide an additional benefit to the client by maintaining an automated system for documents, testimony and other reusable materials relevant to the client's pending and future litigation?
   d. Does the firm maintain a master document repository for institutional clients with repetitive litigation?

 e. Does the firm maintain a master repository of testimony for a client in specialised practice areas, such as product liability, construction or labour litigation?

4. Does the firm plan to use a centralised work product retrieval system?

 a. Will the document management software also be useful for case-specific litigation support?

 b. Will the full-text retrieval software also be useful for case-specific litigation support?

 c. Does the firm have sufficient numbers of personnel trained to index and code work product retrieval documents? Can those employees be trained to code and manage documents for case-specific litigation support?

5. Will an ALSC promote economies of scale in the firm's litigation support systems?

 a. Will an ALSC reduce the costs of repetitive recruitment, by creating a career path for legal assistants?

 b. Will an ALSC reduce the costs of repetitive training?

  (1) Will it reduce the time and expense of training on deposition digesting?

  (2) Will it reduce the time and expense of training on document management, including indexing and abstracting?

  (3) Will it reduce the time and expense of training on work product management?

 c. Will an ALSC reduce delays caused by sending deposition digests to outside service providers?

 d. Will an ALSC better facilitate an even distribution of work among the firm's legal assistants?

 e. Will an ALSC better facilitate an even distribution of work among the firm's attorneys?

 f. Will an ALSC facilitate amortisation of costs over a wider number of cases and/or automation projects?

6. Can the costs of the ALSC, or a portion thereof, appropriately be passed on to the firm's clients as billable services?

 a. Will the handling of the litigation support function in-house be cost-effective for the client?

   b. Will it be cost-effective for the firm?

   c. Can the ALSC be utilised appropriately as a separate profit centre within the firm or the litigation department?

   d. Can the ALSC be administered as a self-supporting cost centre within the firm or the litigation department?

7. Will the ALSC be an effective tool for practice development?

   a. Will the firm be able to handle more complex cases concurrently?

   b. Will the firm be able to offer clients competitive services at reasonable cost?

   c. Will the firm be able to offer institutional clients creative services such as master document indices and testimony databases of recurring company and expert witnesses?

   d. Will the firm be able to offer clients better control over their documentary materials?

## HOW WILL THE ALSC BE STRUCTURED?

1. Can any of the firm's existing systems be utilised for automated litigation support?

   a. Does the firm currently have centralised administration of its computer systems?

      (1) Is there a manager or administrator of the computer systems?

      (2) Is there a manager or administrator of the work product retrieval systems?

      (3) Is there a manager or administrator of the legal assistants?

      (4) Does the firm have in-house technical support or information systems personnel?

   b. Does the firm currently have centralised systems for control of its work product and physical document repositories?

2. Will the ALSC be formally or informally structured?

   a. Will the firm decentralise administration of the litigation support activities?

      (1) Will litigation support activities be spread throughout the firm, utilising multiple stand-alone computers?

      (2) Will each case or project be handled individually by the attorney or trial team involved?

      (3) Will litigation support activities be networked in any way?

   b.  Will the litigation support activities be centralised?

      (1)   Will the firm designate a partner or a committee to oversee the administration of the litigation support activities?

      (2)   Will the firm use a unified ALSC?

      (3)   Will the firm's work product retrieval system be administered as part of the unified ALSC?

   c.  Can the firm utilise existing personnel for litigation support activities?

      (1)   Will legal assistants and word processors with other responsibilities be utilised for litigation support activities?

      (2)   Will firm personnel currently responsible for information systems or technical computer support be utilised for litigation support activities?

      (3)   Will the administrator, office manager or MIS director be given any responsibilities for litigation support activities?

      (4)   Will the firm librarian be given any responsibilities for litigation support activities?

   d.  Will the firm need to recruit additional personnel for automated litigation support activities?

      (1)   Will the firm use permanent or temporary personnel for litigation support?

      (2)   Will the firm use a combination of permanent and temporary personnel for litigation support?

      (3)   Will temporary personnel be employees or independent contractors?

3.  How will the ALSC be physically maintained?

   a.  Will search and retrieval be spread throughout the firm?

      (1)   Will the firm dedicate a single computer for housing litigation support databases? Will it be centrally located for easy access by all attorneys and legal assistants who require access?

      (2)   Will the attorneys and/or legal assistants have individual workstations in their offices?

      (3)   Will the attorney/legal assistant workstations be networked?

   b.  Will document coding and database management activities be dispersed throughout the firm?

      (1)   Can the firm segregate any portion of the library for litigation support activities?

(2)    Can the firm dedicate a conference room or other separate facility for litigation support?

c.  Will a unified ALSC be physically housed in one central location?

(1)    If the firm has multiple branch offices, will the ALSC be in the main office?
(2)    Will branch offices have satellite litigation support centres?
(3)    How will branch offices communicate with the ALSC in the main office for involvement in database design?  For monitoring of coding? For search and retrieval?
(4)    Will the document coding take place in the same location as the document repository is housed?
(5)    Will the document coding take place in the same location as the computer work stations are located?
(6)    Will testimony enhancements and maintenance of full-text databases take place in the same location as the computer database is housed? If not, how will on-line summarising and enhancement take place?

## HOW WILL A CENTRALISED ALSC BE STAFFED?

1.  Will the ALSC use a Litigation Support Manager?

2.  Will the ALSC use project managers for case databases and/or special projects?

3.  Will the ALSC use coding supervisors?

a.  Will coding supervisors be responsible for quality control?
b.  Will coding supervisors be responsible for data entry personnel and quality control of the data entry function?

4.  Will quality control be a separate function?

a.  Will the ALSC use quality control supervisors?
b.  Will the ALSC use quality control specialists [as distinct from coders or Coding Supervisors]?

5.  Who will be responsible for document coding and administration of document repositories for document management systems?

a.  Will the ALSC use legal assistants for document coding and physical document management?
b.  Will the ALSC use separate document coders/analysts?

   c.  Which personnel will be responsible for document coding activities?

      (1)   Will document numbering and physical control be handled by the coding staff?  Who will supervise these activities?

      (2)   Will document coding be conducted on-line, on separate coding sheets or with a combination of approaches?

      (3)   Will the coding staff be trained for both document abstracting and full-text enhancement?

      (4)   Will coding staff ever be responsible for quality control of other coders work?

      (5)   Will the ALSC utilise separate data input personnel when hard copy coding sheets are utilised?

      (6)   Will the coding staff ever perform on-line coding?

      (7)   Who will be responsible for quality control following coding? Following data entry?

   d.  Which personnel will be responsible for document numbering and physical control of the hard copy repository?

      (1)   Will documents be Bates stamped?

      (2)   Will computerised numbering labels be used?

      (3)   Will documents be logged in and out of the document repository?

6.  Which personnel will be responsible for enhancement and database management of full-text retrieval databases?

   a.  Will legal assistants perform on-line summarising of testimony transcripts and lengthy work product documents?

   b.  Will legal assistants perform on-line enhancements of depositions and other textual documents?

   c.  Will document coders/analysts ever be responsible for on-line enhancements?

   d.  Who will be responsible for quality control following enhancements?

7.  Who will be responsible for database design and development?

   a.  Will each trial team perform this function independently?

   b.  Will the firm centralise these functions?

   c.  Will a project manager or other member of the ALSC staff be assigned permanently to each attorney or trial team using ALSC services?

8.  Who will be responsible for developing vocabulary controls coding conventions on each document management database in the ALSC?

   a.  Who will be responsible for development of vocabulary controls, including

issue/subject codes, keyword authority lists and related taxonomies?
b.  Who will prepare the coding manuals?
c.  Who will prepare the user guides?
d.  Who will serve as liaison with the attorneys using the database?

9.  Who will be responsible for developing enhancement/annotation conventions for each full-text database in the ALSC?

a.  Who will be responsible for development of vocabulary controls, subject/issue codes, keywords, and enhancement/annotation techniques?
b.  Who will prepare the coding/enhancements manuals?
c.  Who will prepare the user guides?
d.  Who will serve as liaison with the attorneys using the database?

10.  Will the ALSC staff have any role in search and retrieval on individual databases?

a.  Who will perform search and retrieval on the work product retrieval system?
b.  Who will perform search and retrieval on individual document management databases?
c.  Who will perform search and retrieval on individual full-text databases?

11.  How will the litigation support activities be billed to clients?

a.  Will the firm have an established policy for billing decentralised litigation support activities to clients?

    (1)  Will each supervising attorney be responsible for determining the billing policies for individual cases, projects or clients?
    (2)  Will the firm establish a specific fee structure for billing decentralised litigation support services to the client?

b.  Will the firm have an established policy for billing the services performed by a centralised ALSC to the client?

    (1)  Will the firm establish a specific fee structure for billing ALSC services to clients?
    (2)  Who will be responsible for setting and updating the ALSC fee structure?
    (3)  Who will be responsible for billing the services performed by the ALSC to individual clients? (e.g., the litigation support manager, the supervising partner?)
    (4)  Who will be responsible for reviewing, approving and/or modifying billings by the ALSC to individual clients?

12.    What will be the most effective reporting structure for the ALSC?

a.   Will an individual attorney be designated the managing attorney for the ALSC?  Will this responsibility rotate?
b.   Will a committee composed of attorneys be responsible for managing the ALSC?  Will the committee membership rotate?
c.   Will a legal assistant be designated the manager of the ALSC?
d.   Will the administrator of the work product retrieval system report to the litigation support manager to a committee?
e.   How will personnel responsible for existing firm computer systems, such as time and billing/accounting, be integrated into the litigation support activities?

## HOW WILL ALSC PERSONNEL BE RECRUITED AND SELECTED?

1.   Who will be selected as the **Litigation Support Manager**?

a.   What qualifications are necessary for a successful litigation support manager?

(1)   Does the individual have specialised litigation support experience?
(2)   Does the individual have paralegal experience?
(3)   Does the individual have management and/or administrative experience?
(4)   Does the individual have general litigation budgeting experience? Experience with budgeting automated litigation support projects?
(5)   Does the individual have the confidence of firm's management?
(6)   Does the individual have direct experience with the database software to be used by the firm for full-text retrieval?  For document management?  For work product retrieval?
(7)   Does the individual have experience with the types of databases to be used by the firm?
(8)   Has the individual been employed by the firm long enough to know the firm's practices?
(9)   Does the individual have the necessary skills to perform effectively?

(a)   Ability to exercise independent judgement and assume responsibility in the trial team's absence?
(b)   Communication skills for presentations to clients and/or attorneys?
(c)   Personnel supervision and management of a diverse litigation support staff?

b. What will be the litigation support manager's responsibilities?

(1) To whom will he/she report?
(2) Will the manager formulate ALSC policies?
(3) Will the manager formulate ALSC procedures?
(4) Will the manager work directly with the trial team to develop the database structure on all cases and projects?
(5) Will the manager have any responsibility for approving billings from the ALSC to clients?
(6) Will the manager make presentations to the firm's clients concerning the capabilities of the ALSC?
(7) Will the manager be responsible for direct supervision of project managers? Of quality control specialists? Of document coders/analysts? Of data entry personnel?
(8) Will the manager conduct searches or print reports for attorneys or legal assistants?
(9) Will the manager design and/or conduct training of the litigation support centre staff, including project managers, QC supervisors and/or coders?
(10) Will the manager design and/or conduct training of the attorneys on ALSC policies and procedures? On the use of the software and hardware? On enhancement/annotation techniques? On search and retrieval techniques?
(11) Will the manager prepare litigation support budgets on cases and/or projects handled by the ALSC?
(12) Will the manager be responsible for determining when and under what circumstances outside litigation support service bureaus will be used in whole or in part on a project?

  (a) Will the manager prepare the Request for Proposal or project specifications?
  (b) Will the manager make recommendations to the trial team for retention of external litigation support services.
  (c) Will the manager coordinate or assist in the coordination of litigation support resources between the firm's ALSC and an outside service bureau?

2. Will the firm utilise **Document Coders/Analysts**?

a. To whom will they report?
b. How many document coders/analysts will be assigned to each case?
c. What will be the responsibilities of the document coder/analyst?

(1)    Will they specialise in document management or full-text databases?

(2)    Will they perform on-line document coding?

(3)    Will they perform on-line enhancements and/or document summaries for full-text databases?

(4)    Will they have any responsibility for maintaining the physical document repositories?

(5)    Will they have any responsibilities for the firm's work product retrieval system?

(6)    Will they conduct searches or generate database reports at the request of a project manager?

(7)    Will they have any responsibilities for direct liaison with the trial team?

d.  What qualifications will be necessary for the successful document coder/analyst?

(1)    Will they need specialised substantive experience with particular disciplines, such as medical or scientific terminology, construction, patent or real estate documentation?

(2)    Will they need a college degree?

(3)    Does the prospective coder have the necessary skills to perform the work?

(a)    Is she/he detail oriented?

(b)    Is she/he reliable?

(c)    Can she/he limit coding to objective rather than subjective criteria?

(d)    Can she/he follow directions carefully?

(e)    Does she/he have good language and vocabulary skills?

(f)    Does she/he work effectively under pressure?

(g)    Does she/he have experience with the database software or will training be required?

(h)    If on-line coding or enhancements will be required, does the candidate have sufficient typing skills?

3.  Will the ALSC need independent **Data Entry Personnel**?

a.  What qualifications are necessary?

(1)    Is the individual sufficiently detail oriented?

(2)    Does the individual have adequate typing skills?  Speed?

b.  Will coding and data entry be combined?

## HOW WILL THE ALSC BE ADMINISTERED AS A PROFIT CENTRE?

1. Will clients be billed for the services of the ALSC?

   a. For physical document inventory, document control and number stamping?
   b. For development of the database structure?
   c. For development of the vocabulary controls, taxonomies and keyword authority lists?
   d. For development of issue codes or subject category codes?
   e. For development of coding manuals? For updating of coding manuals?
   f. For development of user guides? For updating of user guides?
   g. For training of the coders, QC supervisors and project managers on the case-specific issues in each litigation support project?
   h. For document coding and analysis on document management databases?
   i. For enhancement of full-text databases?
   j. For QC controls during coding and enhancements?
   k. For data entry and quality control during the data entry process?
   l. For search and retrieval by ALSC staff, when directed by the trial team?
   m. For search and retrieval by attorneys and/or legal assistants?
   n. For preparation of reports and print-outs of search results?

2. How will the fees be set on document management projects?

   a. Will the client be charged on a per hour basis? On a per document basis? On a combination of approaches?
   b. Will the client be charged by the hour for any of the following services:

      (1) Database design.
      (2) Development of vocabulary controls, subject/issue codes, coding conventions, coding sheets.
      (3) Development of report/print-out formats.
      (4) Search and retrieval.

   c. Will QC be charged separately on any basis, such as:

      (1) as an hourly service?
      (2) as a flat fee per page or per document?

   d. Will search and retrieval as part of the QC process be charged separately?
   e. Will data entry be charged separately?
   f. Will the database software facilitate accurate timekeeping for document coding, data entry, search and retrieval and report generation?

3. How will the fees be set on full-text retrieval projects?

   a. Will the client be charged by the hour for any of the following services:

      (1) Database design.
      (2) Development of subject/issue codes, vocabulary controls and enhancement techniques.
      (3) Preparation of enhancement coding manuals and user guides.
      (4) On-line summarising and enhancement.
      (5) Development of report/print-out formats.
      (6) Search and retrieval.

   b. Will QC services be billed separately by the hour?
   c. Will search and retrieval for QC be billed separately?
   d. Will database loading of full-text documents be charged separately?
   e. Will the text retrieval software facilitate accurate time-keeping for billing database loading, on-line enhancements, search and retrieval and report generation?

4. Will the firm charge clients for storage of data on its computer systems specifically:

   a. For document abstracts on a case-specific document management database.
   b. For document abstracts and indices for recurring evidentiary documents relevant to multiple cases for the particular client.
   c. For textual material on a case-specific database.

   d. For textual material on a testimony retrieval system (for recurring client and expert witnesses).
   e. For textual material such as case-specific work product.

5. How will fees be set for ALSC services?

   a. Who will determine the hourly rates for the litigation support manager, project managers, QC personnel and document coders/analysts?
   b. How will the per page fees for document coding be set?

      (1) Who will determine the complexity of the factual and legal issues, and their impact on per document coding and QC fees?
      (2) Who will determine and evaluate the anticipated amount of time for document coding, based on the nature, scope and complexity of the documents?

   c. Will fees for services performed by the ALSC be included as a legal service or an expense item in the disbursement portion of the billing statement?

6. Does the firm have a method for providing a litigation support budget to the client at the outset of a project?

   a. Who will be responsible for preparing the litigation support budget?
   b. How will the attorneys' input be received and incorporated?
   c. Will actual costs be tracked against the budget? Will the budget be revised during the project, as required?

## IS AN ALSC FEASIBLE FOR THE FIRM?

1. Is there sufficient space for a document repository?

   a. Is it centrally located?
   b. If not centrally located, is it accessible to attorneys and legal assistants? Is it accessible to the litigation support centre?
   c. Are hard copies of documents accessible from remote locations? Will the system need a fax board for transmission to remote locations? Will users in remote locations need a printer?
   d. Is there adequate physical security for housing confidential documents?

      (1) If confidential documents must be segregated, will the facility accommodate this?
      (2) What specific security procedures will be required?

2. Is there sufficient space for document coding?

   a. Is the location physically secure?
   b. Is the location secure from a confidentiality point of view?
   c. Is there sufficient space for document coding? For QC?
   d. Is there sufficient space for on-line document coding?
   e. Is there sufficient space for data entry from hard copy coding sheets?

3. Are the attorneys sufficiently committed to using the services of the ALSC?

4. Will recruitment, training and supervision of ALSC personnel be cost effective for the firm?

   a. Is there a sufficient applicant pool for document coders/analysts? For QC specialists? For project managers? For data entry personnel?

b. Will the anticipated work flow of the ALSC permit using temporary or "project" employees for document coding, quality control, repository management and data entry?

c. Does the firm have a stable and competent team of paralegals who can form the basis for the management and search/retrieval/reporting functions of the ALSC?

## HOW WILL EFFECTIVE TRAINING BE ACCOMPLISHED?

1. Who will be responsible for general litigation support training?

   a. For permanent litigation support staff?
   b. For temporary litigation support staff?
   c. For new attorneys, law clerks, legal assistants and others who will be using the system for search and retrieval only?

2. Will training be required on new or updated software systems?

   a. Will the vendor provide training?
   b. Will the litigation support manager provide training?
   c. Will the firm use an outside consultant?  An in-house specialist?
   d. Will periodic refresher training be required for sporadic users?

3. How will new ALSC staff -- coders, QC personnel, data entry operators -- be integrated into the training program?

4. How will new users be integrated into the training program?

5. Will the firm maintain a written training program?

6. Will training be conducted on-site or off-site?  Will it be during business hours? (If not during business hours, will overtime compensation be required?)

7. How will case-specific training be accomplished?

   a. How will the attorneys' input be solicited and received?

      (1)    On factual/legal issues.
      (2)    On required vocabulary controls.
      (3)    On report formatting.
      (4)    On changes during discovery?

b. How will vocabulary controls, coding conventions, enhancement techniques and other case-specific issues be addressed in training?

    (1)    How will these be integrated into the coding manual?
    (2)    How will these be integrated into the user guides?

c. As case-specific vocabulary controls and coding conventions change during discovery, who will conduct ongoing training updates?

    (1)    Will these be written, verbal or both?
    (2)    How and when will these be integrated into the coding manual, user guides and written project materials?
    (3)    Who will determine when the case-specific issues, vocabulary controls etc. have changed sufficiently to require formal training updates?

8. How will attorneys and other members of the trial team be trained on search and retrieval and enhancement techniques?

a. Will users in branch offices require training? Where and when will it be conducted?
b. Will co-counsel require special training and security passwords for access to any databases during discovery and/or trial?
c. Will client representatives require special training and security passwords for access to any databases during discovery and/or trial?
d. Will any consultant or expert witness working with the trial team require special training for access to any part of the database?

# 11. Document Image Processing the First Step .... and Beyond

*Jimmy Mackintosh*
*Lawrence Graham*

No doubt you have been convinced by the presentations which you have already heard that document image processing is for you and your firm. No other technical approach can realistically cope with the fact that 95% of all the information going round your organisation is in paper form. No other technology seems to offer the promise of regaining control over the mass (or mess?) of documents which otherwise remain inaccessible or accessible only at a disproportionate cost.

On the face of it, your clients would have no beef with you if you turned your organisation fully over into paperlessness. They could expect a faster turnaround of the work, an even higher quality of work product and would no doubt expect to pay a larger amount for this enhanced service.

From your point of view, document management has become rather glamorous and no longer a drudge. Your data have become usable information instead of inert occupiers of floor space. You are now back in charge of that case which threatened to swamp both you and your department and you can even go out into the marketplace to attract new major litigation cases and come up trumps in competitive tenders. Suddenly recruitment has become a doddle — all the best talent is beating a path to your door. And you know that you have only scratched the surface of the potential benefits — experience has taught you that there are many benefits which you could not have foreseen when you took the plunge. They have only become obvious with the gift of hindsight.

Well, what are you doing still sitting there and not rushing out to the exhibitors, cheque book in hand.

Realistically, there are no doubt a number of reasons why there is still an audience for this seminar. In my view, the chief ones are:

*V. Mital (ed.), Advanced Litigation Support & Document Imaging, 127–135.*
© 1995 *UNICOM Seminars. Printed in the Netherlands.*

- the cost of the technology - no doubt dropping fast along with all hardware costs but still well beyond the petty cash threshold

- the possible transience of the problem in relation to the size of the investment — some might say that if the business community took up the use of Electronic Data Interchange at the appropriate rate, the paper mountain would be under control anyway.  Others might well say that false dawns are two a penny in the field of technology

- the hugeness of the organizational and cultural changes which a proper implementation of DIP technology would entail

- the newness of the technology, leading to the typical characteristics of the bleeding edge

- lack of accepted methodologies

- lack of accepted standards

- lack of skilled and experienced help in our market

- a questionmark over the adequacy of your existing staff to cope with the challenges involved

In a word, there are pups out there which someone is going to buy and you don't really fancy that person being you.

Moreover, the characteristics of the law firm mentality may not be conducive to a major implementation of DIP at this stage.  Risk aversion, a lack of strategic investment and a timesheet mentality all contribute to the barriers which are erected.  Add to that anecdotal evidence that not all DIP installations have been without their problems, both technical and operational, and it is hardly surprising that those in charge of the purse strings are not rushing to loosen them without much persuasion.

In these circumstances, everything points to a softly, softly approach.  In the legal profession we are fortunate in that we can isolate, relatively easily, a manageable area of our work for the purposes of running a pilot study.  While a bank or an insurance company would have to plot its course far into the future before even an experimental installation could be made, it is easy for the law firm to choose an individual or a group or a single file on which to experiment.  Instead of major mainframe systems development or purchase, we may be talking about using existing hardware and pricing determined by the PC marketplace.  No longer do you have to make sacrifices in terms of power and functionality if you do use the PC platform for your pilot: even multiuser pilots on a fully specified basis are possible thanks to the availability of powerful networks.

Before embarking on our pilot, we must notice in passing that we have already breached one of the golden rules of law-firm management, at least as she was spoken in the 1980s. The information technology strategy exists in order to support the business plan. But here we are seeing an example of the very availability of the technology leading to a significant alteration in the way we conduct a section of our business - technology driving the business strategy. I stress the need for management to be as technologically aware as they can possibly be, because this is only one area in which the legal profession is going to be faced with new opportunities and challenges thanks to developments in technology.

## 11.1   THE PILOT

Rule 1: choose your pilot project carefully. There are a number of factors to take into account in deciding which study is likely to yield the best results. There are plenty of potential areas for a pilot, e.g.

- a currently running litigation file where control over the documentation is already proving to be problematical

- the litigation file which you know is about to walk in through the front door where you can foresee a heavy document content

- departmental knowhow databases

- the centralised information bank

- the current filing requirements of any work group or individual

- a section of the file archive for a section or work-group.

We do of course have to define what it is that we want out of the pilot. We would not be embarking on it at all if we did not have confidence in the theoretical justification for it, and there are enough statistics floating around to sink a battleship. The main reasons for a pilot must therefore be:

- to make sure that the particular manifestation of the technology which some persuasive salesman has foisted on you actually matches up to the theory

- that your organization can handle the cultural changes which are inevitably involved

- that the pilot's success will convince the doubting Thomases who hold the pursestrings in every firm that the technology actually delivers.

So - Rule 2 - try to find a project that actually makes money. A pilot that merely saves money is a second class citizen in the eyes of the unconverted — infinite room for argument over whether we needed to spend the money in the first place. That probably puts the project squarely in the litigation department.

Rule 3: make the project as high profile as you dare. Indeed unless you feel reasonably confident that you want the project to be high profile you may not be quite ready to undertake it.

This in turn means that you have to have all the elements of good project management in place, so let's have a look at the key features there.


## 11.2    THE PEOPLE

Every successful project requires the wholesale commitment of the implementation team and of the line management. The backing of the boss is vital because every project is going to go through rough patches. We are planting an acorn which is going to grow into a mighty oak but in its first few months of life, it is going to have a fragile existence. Someone has to be there to protect it from damage, malicious or natural, to make sure that it is given sufficient sustenance and otherwise to see that the team of arboriculturists are doing all the right things.

As far as the team is concerned, it is vital that they "own" the project and that this is not another of those adventures which the IT department insists is undertaken which has little relevance to the way that we really do our business. The problem here is that all good IT projects up to now have tended to take as the model the way that we conduct our business currently — don't antagonise the users by asking them to change the way they work: rather, get the systems to mimic the manual methods in order to smooth the transition. Not so easy where the real benefits can only be fully realised by changing some of the underlying processes involved in doing the job.

This is not a seminar about computerised litigation support, but to explain my point, let us look at the processes which may be introduced alongside the implementation of an image system in that area of a practice.

- the arrival of an IT aware specialist who is going to tell you how he wants you to organise your filing

- the arrival of paralegals or possibly outside contractors who are going to code up your documents in a way on which you are having to rely in accessing those documents in future

- working off a screen rather than with piles of comforting paper

- sharing work on the file with other workers in the team and therefore having to record your comments on documents at the keyboard

- dealing with discovery of documents through the machine

- developing the mind set which can cope with Boolean operators as the chief means of negotiating your way round the files.

By no means an easy switch to achieve and it is clear that there must be considerable amounts of time available for the fee earners to be "reprogrammed" in order to be able to cope with the new methods. That in turn means that there must be sufficient resource in the IT section to cope with day to day support, even after the initial training phase is completed.

That all speaks to me of the mistake it would be if you tried to introduce image processing to a running case. Nice as it would be to offer immediate benefits to hard-pressed fee earners, I believe that it would be almost impossible to achieve that allocation of time and resource by both the users and the supporters to give the pilot a reasonable chance of success. The first panic that came along where the new system could not compete with the old without further development or training and it would take more than the average commitment from senior management to protect the young sapling.

Ideally, you will get back to the office later this week to find a message from your favourite client that he wishes to pay for you to introduce an image system to handle a new case for him. Alternatively, we may have to wait a long time for that happy day to arrive and the question we must ask ourselves in the meantime is will that client come along in a year's time and say the same thing or will he have gone down the road to Messrs Jones & Co because they can offer a fully functioning, tried and tested system. It is clear that there are clients who are aware of the technology and will expect the firms with whom they place their work to be up with the hunt in this area.

## 11.3   RESOURCES

There are real questions of expectation management in piloting DIP installation. Users are likely to have read about how far one can go in creating the paperless office and may well assume that next Monday they can wave goodbye to the piles of files which currently grace their rooms. You, on the other hand, have budget approval for PC, a scanner, some OCR software, a full-text search package and an optical drive; and a list of other tasks of equal urgency which means that you have precisely one day to get the whole project underway. Save yourself the trouble and give your budget back just as quickly as you can, for, however enthusiastic the fee earner, the pilot will fail unless the appropriate resources are devoted to it.

Admittedly, you are going to have to come clean on the real costs involved, in terms not only of hardware and software but, more importantly, "liveware", but in the long term it is better to do nothing at all until the project can be adequately resourced rather than going off at half cock and playing into the hands of the "I told you so" brigade.

This is all part of the preplanning stage which will take many man days if it is to be done adequately. The debate will include whether the project is to be conducted purely on internal resources which may mean immediate recruitment, or whether it is safer to rely on external consultancy help.

If the latter, are there consultants out there who not only understand the underlying technology in appropriate detail but understand the processes and culture of a law practice? With a newish technology, the answer is that even if you do use external consultancy, your firm must rapidly develop the home grown skills to transfer the project in-house quickly and to provide supervision of what the consultants are up to. Bringing in consultants is no substitute for the hard work of coming to terms with the technology and the human resources implications of its implementation but may be a means of getting the project bump started.

I cannot stress too much the need to grasp this opportunity to look at the processes involved in conducting a piece of litigation rather than merely automating existing tasks. That is where the full benefits will lie although admittedly the up front disruption and change management is more far-reaching and difficult to handle.

## 11.4   INFRASTRUCTURE

With any luck, the basic technology may already be in place in your firm to be able to run a pilot without significant departure from the building blocks already there. The single user can find out the capabilities of image systems running on a PC with associated kit such as scanner, printer, etc. The team worker for whom the pilot must involve multiple access will find appropriate technology either in the network or the UNIX marketplace. Certainly prudence would dictate sticking to standard kit wherever possible and there seem to be no reasons why this cannot be achieved without sacrificing functionality or performance.

This is just one of many areas where the legal profession seems to find that the world of technology is catching up with its needs. From the obviously back-office applications such as accounting packages and word processing, we are now witnessing an explosion of potential front office applications: image processing, computer assisted drafting, on-line databases, CD-ROM publishing, workflow software, hypertext, speech recognition. In order to keep in the mainstream of these exciting developments, it does seem to me that the safest course does lie in the DOS/networks infrastructure.

A point to watch is the capacity of whatever network you are using. Whenever you are moving image around a network, you are bound to be using significantly more capacity than moving text. Ultimately this will point to the need for a high capacity network based on the fibre optic standard (though how far one can yet say that it is a standard is debatable) and that will be more expensive than ethernet or token ring. But as you look forward to other image movements around the network, e.g. fax to the fee earner's desk, it is as well to build in an upgrade path to your existing network configuration.

A personal opinion is that a graphical user interface is a considerable advantage for whatever software you are going to ask fee earners to use. This is chiefly because of the ease of training and indeed cross training from one software package to another but also because of standard tools allowing two or more packages to transfer information easily between them. Some would say that the Microsoft Windows operating environment is not sufficiently robust to support "serious" and mission critical operations, but the view seems to be that version 3.1 has sorted out the major resiliency problems.

Clearly there will be counter arguments, particularly if you already have a different system and your supplier can offer an alternative package. Which way you jump will also depend on how far and how fast you can go - can you support a multiuser pilot? Is the pilot in respect of an individual, departmental or centralised application? What response times and capacities do you require? However, while these factors would even in recent times have ruled out solutions built around the humble PC, nowadays the increasing power and capacity of the kit in the PC marketplace has put the PC in contention in virtually every office application.

## 11.5   WIDER IMPLICATIONS

Bear in mind the directions in which this is all leading you. A successfully implemented image application will not stop there. The hope is that the benefits derived from the experience will open up new areas and, longer term, the end result ought to be the near paperless office. If that is the direction in which we are indeed headed, there are a few minor matters which must be sorted out en route!

I am told by people who understand these things that computer-held images are not acceptable as evidence under the Civil Evidence Act 1968 and the status of text which has passed through an image stage on its way to ASCII text may be open to doubt.

What of documents which still require to be "in writing" as a matter of law?

We still need to address the issue of the ease in which data in digital form can be manipulated in ways that are to all intents and purposes untraceable. Authentication by means of electronic signature, while technically possible, needs agreed standards.

Are we satisfied that we are safe from interference from viruses or from accidental or malicious damage? What will happen in the next power strike? Have we dealt in society with data protection issues?

Subject to those points, we can look forward to possibilities opening up which could transform the way we work, if our working materials (filing cabinets full of files, bookshelves full of text books, etc.) can all be reduced to a small set of easily transportable disks. Will we really want to continue with the delights of commuting into the major conurbations if that is no longer necessary? Will firms wish to continue to pay central city rents when there is no need to bring staff to one central point to enjoy a pooled infrastructure?

Perhaps we will see the resurgence of the sole practitioner or small practice linking up with others of that ilk in order to provide wide coverage without having to work for a monolithic firm. I can envisage the growth of support organisations which exist to provide support for such networks of individuals and small firms, providing not only technical support but also materials such as precedents (in a form to be used with computer-aided drafting packages) knowhow and even text books (in machine readable and hypertext form).

Closer to home in terms of time, it is encouraging to see the lead which is now being given by the judiciary. It will not be long, I suspect, when there will be real pressure from the judiciary for the parties to agree to use a particular DIP approach, if not actually a particular package, just as there is currently pressure to adopt particular word processing and communications conventions. As judges see the practical benefit of computer support, e.g. the pilot study of computer aided transcription in the Official Referee's court, the pressure for change from the top will increase.

At present, imaging is probably perceived in the courts as being applicable only in the huge commercial cases, but with the falling cost of the kit and increasing awareness of the benefits of control of documentation, the trickle down into all areas of practice, cases both large and small, will not be slow in coming.

That is important from the point of view of maintaining the UK's position as a centre for international dispute resolution, when the client base is increasingly aware of what firepower is available for deployment on his side and will not accept second best.

## 11.6　CONCLUSION

I believe that we are now emerging from the frontiersman stage in the evolutionary process of DIP technology. To have gone ahead even with a pilot study in 1990 would have required a long purse, infinite patience and strong nerves. No doubt, those who blazed a trail well deserve the start that they have achieved. It is thanks to them that we, the second wave, can reap the benefits of having a route map through the minefield. If we want to respond positively to the challenges of doing business as lawyers in the 1990s, this is certainly one technology with which we must come rapidly to terms and I believe that we will enjoy doing so.

# 12. Successful Investment in Litigation Systems

*Clive Thorne*
*Denton Hall Burgin & Warrens*

## 12.1 FACTORS IN THE DECISION TO INVEST

### 12.1.1 Identification of Need

My firm's decision to invest in a litigation support system was a result of a unanimous decision by all the litigation partners within every area of practice. The advantages were, however, most clear in a number of areas including the following:

- "big case" litigation

- "specialist" litigation, such as construction litigation, intellectual property, damages inquiries, product piracy campaigns and fraud

- "rapid" litigation, such as interlocutory injunctive work

- arbitration.

### 12.1.2 Competition and Cost

Behind the practical needs to develop litigation support was the ever-present factor of cost. We prepared a number of permutations and projections as to the way in which a word retrieval system could assist in a major discovery. It was manifest that the support system would reduce the fee earner cost and enable the saving to be passed on to the client as well as, in any event, enable fee earners to function more efficiently. It should be remembered that when in 1987 we considered this issue fee earner costs were rising dramatically.

We had also seen the benefit of the debt collection package which was used by our office in Milton Keynes and had learned of similar experiences from firms which had used such packages in possession actions. There was also a feeling that correct

*V. Mital (ed.), Advanced Litigation Support & Document Imaging*, 137–141.
© 1995 *UNICOM Seminars. Printed in the Netherlands.*

use of a support system would provide opportunities for much "neater" and organised work. This would be attractive to clients, whether new clients, target clients or existing clients.

There was very little example of litigation support systems being used by major firms in England. It was therefore necessary to undertake a fact-finding and evaluation exercise in the United States. In his paper, Howard Field looks at the detail of this exercise.

My firm relies heavily upon international clientele and particularly clientele, professional and otherwise from the United States. It seemed that if we were able to take the best from the US systems and procedures that we saw, then this in itself would be attractive to many overseas (as well as domestic) clients.

We were also conscious of the fact that London, although an expensive jurisdiction was the litigation capital for Europe, that there was general overseas confidence in the UK civil litigation process for major cases and that there was therefore no alternative to develop such a system if London was to retain this pre-eminence.

### 12.1.3 Marketing Pressures

I have the luxury of practice in a specialist area of litigation which enables me to promote my practice more readily. I have often wondered how to promote effectively general commercial litigation. The development of efficient litigation support systems is one way in which a litigation practice can be marketed. Perhaps the non-lawyers in the audience will forgive me when I suggest that a system of litigation support may persuade a client to litigate when he is otherwise fearful of doing so. Certainly our experience has shown greater cost-effectiveness in the conduct of litigation, a point which can justifiably be put to clients in assessing the benefits or otherwise of commencing an action.

### 12.1.4 A Vision of the Future

Overall we felt that the future for the conduct of litigation required firms to maximise their use of legal technology. In the same way that lawyers in the 1970s adapted (perhaps too eagerly) to the facility for the mass copying of documents and in the 1980s to the fax and word processor so, in the 1990s lawyers will, I am sure, accept that their practice is going to depend upon being linked to a database whether in their offices, at home, on the client's premises, in counsel's chambers or in the court room.

## 12.2   LEARNING FROM THE US EXPERIENCE

It was quite clear (see Howard Field's paper) that the United States had a headstart in the development of litigation support systems, whether using PC based software or on a more substantial basis.   The evaluation of the US experience was a fascinating exercise.  We were greatly assisted by Patti Eyres who I am delighted to say is attending this seminar.  The more exciting part of the programme was the training of the litigation support group and the formation of a system.  Here we relied heavily upon an experienced litigation support manager, Lisa Mounteer, who had wide experience in the United States.  The short time it took from delivery of the software until the time the system was up and running was largely due to the lessons that we learned from the US experience as well as the contributions from consultant Patti Eyres, and Lisa Mounteer.

## 12.3   THE DIFFERENCE BETWEEN US AND ENGLISH PRACTICE

This sub-heading can perhaps be translated into the concept of accepted practice v. novelty.  In the United States litigation support was accepted.  In England it was novel and in 1987 and 1988 treated with some scepticism.

There are a number of considerations which are highly relevant.  Not the least of these are the different procedural needs.  For example, pre-action discovery is a common step in the United States.  Great emphasis is also placed upon written depositions.  In England, by contrast, a great deal of emphasis is placed upon preparation for the eventual trial of an action and the preparation of proofs of evidence and witness statements for the trial.  Thus the inter-relationship between discovery and the preparation of a witness's evidence is a crucial role for litigation support in England.  More so because of the recent introduction of Order 38 rule 2A RSC which now requires the pre-trial exchange of witness statements.

Another factor which distinguishes the United States from England is the split profession.  I have always thought that barristers have been more advanced than solicitors in their use of the word processor for the preparation of opinions and pleadings.  Perhaps this is due to the fact that barristers are more self-reliant when it comes to secretarial services.  Nevertheless, an important object for litigation support is to facilitate the transmission of information between the barristers and solicitors running a case.  This is not so in the United States where litigation support is normally directed within a particular law firm.

The conservatism of the UK court system is changing.  There are a number of courts being built (e.g. the new Courts block in Chancery Lane) which is geared to the use of computer-based technology during the course of the trial so that the judge, barristers and counsel in the case all have access to the same database of documents. Nevertheless, progress has been slow. Although the Lord Chancellor's Department has shown willingness to assist practitioners in the use of computer-based support, there is invariably a financial constraint. Further, many court rooms are much as they were when they were constructed in the 19th century.

The influence of the use of office technology by barristers in the 1970s and 1980s is now beginning to manifest itself on the bench. There have been very encouraging signals from many members of the judiciary with regard to the use of technology during the course of a hearing as well as the use of the products of litigation support, e.g. graphics and schedules during the course of a trial. One is particularly grateful to the path-finding work of judges such as Mr. Justice MacKinnon in the recent Blue Arrow fraud trials and Judge Forbes in the Official Referees Court.

## 12.4    THE RESULT - SUCCESS?

I am in no doubt that the introduction of a litigation support group into my firm's litigation department has been a success. Could it, however, have been a greater success? Without wishing to appear arrogant, I think yes. By that I mean that we haven't yet exploited the benefits of the system to their fullest extent. The reason why not is largely due to extrinsic factors which I have already hinted at.

For example, it would be helpful as a matter of course if all barristers and solicitors acting for other parties as well as the judiciary involved in the hearing of an action were able to link up to the system.

Despite a considerable effort in informing clients of the facility, one still encounters suspicion and concern that the system is going to increase costs rather than decrease costs. I am sure that it will take another five years or so before clients accept litigation support as a matter of course. On the other hand there can be no doubt that the litigation support system has given us the following capabilities:

- much greater facility and speed in the preparation of cases

- the ability to conduct document and evidence analysis

- the preparation of graphics, accounts and schedules of a high quality

- greater cost-effectiveness and time-saving in the conduct of litigation

- increased client acceptance

- bar acceptance

- court acceptance

- fee-earner acceptance

- profitable use by solicitors outside litigation practice

- imaging of documents.


## 12.5　THE FUTURE

It may be that those firms with a developed litigation system will create a litigation super-league with the capability of handling efficiently the largest litigation. On the other hand there are many systems available which increase the capability of firms with small litigation practices. I have already mentioned that part of the prompting to develop a system came from the example of the early debt collecting systems.

Perhaps after all, the litigation support system is no more than a tool in the hands of firms which already have the capacity to conduct major litigation. What is, however, abundantly clear, is that the conduct of litigation has begun to change and will change increasingly rapidly as a result of the development of more sophisticated system, greater acceptance of the use of such systems, greater training of the fee earners and the ability of fee earners to use that system. It is perhaps no coincidence that it is the youngest fee earners who seem to have developed a greater flexibility in the use of systems.

# 13. Setting up an In-House Capability

*Howard Field*
*Denton Hall Burgin & Warrens*

### 13.1.1 Analysing the Firm's Needs

There are two ways in which it is possible to analyse your own firm's needs. The first is by becoming involved in a large case with a considerable volume of documentation. This will compel you to consider what options are available to you in order to manage the case. This is what happened at Denton Hall in 1987 when we were acting on behalf of an American insurance company in relation to the PCW affair at Lloyds. At that time we had over a million documents in the office and at one time had a team of 38 people working on the documentation. As I was responsible for administration of the Litigation Department at that time I was given the task of investigating how we might in the future be able to overcome some of the obvious problems that we encountered.

The second way is for you to analyse your firm's needs and how technology can be used. You can then consider starting in a small way with a PC based software programme. You can then develop your progress as skills increase. At a later stage you may want to establish a Litigation Support Group.

### 13.1.2 Technical Expertise

In 1987 I commenced to investigate what systems were available in England and to see what expertise was available. You will not be surprised to learn that there was little available in this country. In these circumstances I decided to visit America in 1988 to talk with a number of American firms where we had contacts who had existing Litigation Support Departments in operation. I was also able in the summer of 1988 to make contact with Patti Eyres who is here today who became our technical consultant on the establishment of a Litigation Support Department at Denton Hall.

When I went to the USA I had help from large firms in New York such as White & Case and Sherman & Sterling. On the West Coast I was able to talk with

*V. Mital (ed.), Advanced Litigation Support & Document Imaging, 143–148.*
© *1995 UNICOM Seminars. Printed in the Netherlands.*

Jones Day Reavis & Pogue with whom our Hong Kong office was working in close collaboration on a large case in Los Angeles using litigation support. I was also able to visit Seltzer Caplan Wilkins & McMahon in San Diego.

The great difficulty initially was trying to find software which was capable of managing a large case. Having visited the USA I found that quite apart from big case management there were a large number of PC based programs available which were being used in different ways by American attorneys.

On my return to London we undertook the testing of a program developed in the USA which would run on our Wang System. Although the program had a number of useful features it was not fully supported in technical terms in England. We spent a considerable period of time talking on the telephone to the vendors in California to overcome problems. We ultimately decided that without full technical support there was little point in persevering. During 1989 we were introduced by Wang to a software program known as Mires which is a sophisticated information retrieval system. Patti Eyres was able to spend a week in Phoenix assessing the program. Following a series of discussions with Haessler Systems, the German company who produce and develop Mires, we decided to proceed with the establishment of a Litigation Support Department. We have had considerable assistance from Haessler in developing Mires for litigation support purposes. Most importantly we were assured of technical support from both Wang and Haessler themselves.

### 13.1.3 Methodological Expertise

We again decided, as there was little expertise in this country available, to recruit someone from the United States to come and manage our Litigation Support Department. We therefore advertised on both the East and West coasts of America. There was a very good response to our advertisement with a number of highly qualified applicants applying. We eventually interviewed a short-list of eight applicants at Long Beach in California. As a result of this we appointed Lisa Mounteer who arrived in this country in July of 1990 to take up her position as our Litigation Support Manager. Lisa had at that time had over seven years experience of litigation support. She had also had the added advantage of having established the Litigation Support Department of Sidley & Austin in Los Angeles and of running that Department for a period of three years.

With the arrival of Lisa Mounteer we started to address other areas in which the Litigation Support Group could help the Litigation Department not only in the management of large cases but in other ways. The result is that our Litigation Department now is able to handle the following work:

- computerised document indexing

- computerised fulltext searching of court transcripts

- computerised data comparisons, organisations and lists

- damages assessments or calculations using spreadsheets

- graphics illustrations for data for court or analysis.

### 13.1.4 Composition of a Litigation Team

Obviously the composition of a litigation team will depend on the size and nature of the litigation concerned. There must inevitably be one partner who has overall responsibility. In a very large case he may be supported by other partners. Depending on the issues in the litigation the tasks involved may divide up easily or may be shared. In a large case there will be a number of other fee earners involved. A typical sort of case may involve two solicitors with five years experience and there may be four newly qualified solicitors involved together with a number of trainee solicitors.

If the Litigation Support Department becomes involved then it is necessary to assess the number of staff required to cope with the tasks involved and the time within which the work needs to be undertaken.

### 13.1.5 New Management Structures

At Denton Hall the whole firm is structured on a corporate basis using work groups within each department. Litigation Support has become a work group in its own right within the Litigation Department. Whilst I am responsible for the Litigation Support Department the day to day running of the operation is undertaken by Lisa Mounteer as the manager. She has the assistance of a project manager who operates as her deputy. To assist the project manager we have four project leaders. These are staff we have trained and they are able to undertake any of the many different tasks that are involved in litigation support involving the use of a number of PC based programmes as well as Mires.

Since we established the Litigation Support Group we have trained a number of temporary staff who we employ particularly to undertake the task of coding documents. Litigation Support needs very often to react very quickly to requests for help from solicitors. We are therefore able at very short notice and that means within a matter of hours to put together a team of anything between 10 and 20 people to work, sometimes overnight. Our offices are open 24 hours a day and our word processing department is structured so that we have the benefit of three shifts per day. In the Litigation Support Department we frequently work two shifts per day and have the ability, if pressure is very great, of working three shifts per day.

## 13.2    TRAINING LITIGATORS AND SUPPORT STAFF

### 13.2.1  Who can be Trained and for What?

Litigation support staff need to be trained initially on how to code documents. We have found that once they have acquired basic skills that they can then be trained to undertake quality control work and to be trained in the use of the various PC based programs that we use. These programs are particularly R base, Discovery, Excel and Harvard Graphics.

If you choose to employ temporary staff then they will need to be trained for whatever function you wish to use them. We have found that some temporary staff may be particularly experienced in using PC based programs where others acquire the skills of coding and quality control. We therefore assess their progress and focus on their specific strengths and weaknesses in order to ensure that they are properly trained.

The other aspect of training obviously relates to the fee earners who are involved in the case in question. Although we have taught a number of our fee earners basic search techniques we have found from experience that it is quicker and cheaper to use litigation support staff to undertake the searches required by the fee earners. Litigation support is a service department which we have established to respond as quickly as possible to requests for help from fee earners. Searching techniques can take years of experience and we have found from experience that the coding staff who have worked with all of the documents are very often much better placed to be able to obtain the information required. You will appreciate that unless you have a windows based program, then before fee earners can undertake searches on their own they need to acquire keyboard skills. Unfortunately there are a number of fee earners who do not wish to use a computer in any shape or form. Whilst we are able to encourage them and to provide facilities for training we have found it to be counter-productive to try and force fee earners to use a keyboard and screen. On the other hand we have found a number of partners and fee earners who have become enthusiastic and want to acquire skills. As more and more fee earners become involved in the use of litigation support we shall find them quite naturally acquiring the skills that they need without any further encouragement. If you are able to persuade partners to use either a PC or a lap top computer and to lead the way with other staff this will give you a significant advantage.

### 13.2.2  How Long does it Take?

The training of members of the Litigation Support Group in the art of coding will basically take an average of two weeks. To acquire experience in relation to PC based programs will obviously take a further period of time depending on whether the system is being used on a regular basis. If the software is being used on a regular basis then skills can be acquired quite quickly. At Denton Hall we have always viewed litigation support as being 25% software and 75% project management. This means that know-how forms a very significant proportion of litigation support.

### 13.2.3 Formulating a Methodical Programme

Although I have mentioned earlier that we have trained some of our fee earners in search techniques, I do not believe that a structured programme of training will necessarily provide a cost effective way of training staff.

### 13.2.4 What a Self Learner can be Taught

The self learner must both in litigation support terms and from the fee earners perspective have basic instruction on how a typical database is constructed and how the different methods of searching are employed in seeking information. The techniques for full text searching are different from the techniques for coded text searching.

### 13.2.5 Training of a Live Case

There is no doubt that the best form of training for fee earners is to be involved on a live case. The searches which they then undertake acquire far greater significance and they will be able to follow more easily the way in which the databases are designed and the results that can be achieved. We have found that once lawyers have started to use a Litigation Support System they rarely wish to revert to using manual methods.

### 13.2.6 Economic Point of View

As I have already mentioned earlier from a time management and economic point of view it will very often be cheaper from the client's point of view for litigation support to undertake searching and the provision of hard copies of the relevant documents rather than to leave the searching to fee earners who are obviously being charged out at a far higher rate than the staff in litigation support. I believe that the savings achieved by the use of litigation support are approximately one third of the cost of managing a large case on a conventional basis. It also has the additional saving that your fee earners will be fully engaged in fee earning work rather than being involved in the physical management of paper as has occurred only too often in the past.

During the second annual conference on Litigation Support held on 28th March 1992 by the Society for Computers & Law Mr Justice Brooke stated: "Any client who goes to law wishes to obtain the best practicable result. But a lawsuit is, like everything else of importance in his life, an economic activity, and he is entitled to expect that it is conducted for him by the experts he chooses as economically and efficiently as is compatible with the achievement of his aims. He is entitled, in my view, to expect this quality of service not only from his lawyers but also from the tribunal, whether it is a publicly financed court or a privately financed arbitrator, to which he is eventually entrusting his dispute for decision".

# 14. Workflow and Case Management Applications

*Neil Cameron*
*KPMG Management Consulting*

"When I use a word", Humpty Dumpty said, in rather a scornful tone, "it means just what I choose it to mean - neither more nor less".
"The question is", said Alice, "whether you can make words mean so many different things".
"The question is", said Humpty Dumpty, "which is to be master - that's all".
...
"That's a great deal to make one word mean", Alice said in a thoughtful tone.
"When I make a word do a lot of work like that", said Humpty Dumpty, "I always pay it extra".[1]

## 14.1   INTRODUCTION

My brief in this session is to cover two areas which need necessarily have nothing to do with the topic of this conference: imaging. However, just as case management and workflow when combined gain a potential more than the sum of the parts the same thing happens when they are both combined with imaging technology.

I propose to discuss each element in turn and then cover the benefits which can be obtained when all these applications are combined.

As Humpty Dumpty and Alice observe words can mean different things to different people;  and these three concepts can have alternate meanings dependent on the context and the perspective of the user.  Any of you who have witnessed the Pinteresque dialogue resulting from a conversation on the subject of "litigation support" between four people when one person thinks this means "forensic accounting", another thinks of "document indexing", the third "courtroom graphics" and the last "imaging" will know what I mean.  I hope I will be forgiven therefore, in the interest of clarity, when I begin each subject with a discussion of what I think I mean by these terms.

---

[1]   Lewis Carroll, Through The Looking Glass.

*V. Mital (ed.), Advanced Litigation Support & Document Imaging*, 149–158.
© 1995 *UNICOM Seminars. Printed in the Netherlands.*

## 14.2    CASE MANAGEMENT

Case management may be any system that provides "case-holders", whether they be lawyers, social workers or doctors with support in the administration and management of individual "cases".   Such a system for a lawyer would, for example, maintain details of all his matters and clients, together with additional information that may be useful during the life of the case.   Some of this information will be the same as that which one would expect to maintain in a traditional law office recording and billing system, such as:

• client name

• client address and other details

• matter description

• partner and fee earners working on the case.

Other information which needs to be maintained in a case management system will repeat names, addresses and phone numbers that some firms will be holding in a client/contact or conflict of interest database and which would be related to individual cases, for example:

• other parties

• foreign lawyers

• counsel.

Just to complicate things further, a good case management system will also manage all case related documents prepared internally allowing them to be indexed against each case, thus performing the function of a document management system. A point I will return to later when dealing with imaging is the possibility of including in this index the image of all externally created documents, thus maintaining a complete record of all relevant case papers.

In addition to these facilities one would expect a case management system to be able to hold basic milestone information related to the expected life-cycle of the matter, to indicate in respect of each case the last action taken and the next step to be undertaken together with an expected date.   This provides a facility which not only assists the individual case holders in the prosecution of all their cases, but also provides invaluable information on case progress for managers and for clients as well as performance metrics on individual fee earners.

From this it is apparent that a fully featured case management system does not so much co-reside in a law office IT system as much as subsume and replace much of the functionality of existing systems. It is difficult to imagine that the exercise of "bolting on" an additional case management layer on top of existing time recording and billing, client/contact database and document management systems would be successful unless they were designed to work together from scratch. Rather, the best way of implementing such a system is by designing it as a whole around a suitable Relational Database Management System (RDMS) such as Informix, Ingres, Sybase and such like. Fortunately, for this very reason, many of the traditional "Law Society Recognised" suppliers have been developing new RDMS-based systems with case management facilities.

A good example of a case management system, and one of the first legal systems developed, is the Datix system which last year won the Society for Computers and Law Software Award. Datix is an Informix-based UNIX case management system that was developed by Brian Capstick, a solicitor in private practice, who produced the system to manage his own case load of personal injury claims: such systems coming into their own when applied to low value/high volume work. Datix maintains all the necessary case related data as described and provides the case-holder with the ability to monitor, review and progress a large number of cases always with all the relevant facts at her fingertips. The dates assigned to the "next steps" information is useful in reminding the user about deadlines and acts akin to a "to-do" list.

As we have noted, such a system relies on tight integration between the different parts of the system, such integration and interoperability provide other facilities such as the ability to:

- pass automatically the name and address of all addressees to the word processor

- prepare automatic drafts of relevant documents for the "next step" to include all appropriate data from the system.

Another such system under development is the CRIMSOFT system designed in association with the Law Society as a system for the case management of criminal legal aid cases on the basis that the only way to make any money out of such work is to automate as much of it as possible and allow lawyers to handle many more concurrent matters than at present. The CRIMSOFT system is still under development by General Automation.

To summarise pure case management systems from the fee-earners point of view, what he is provided with on a daily basis is:

- easy direct access to all the information he needs to assist in his current matters

- constant reminders of what needs to be done

- assistance with automation of some routine tasks.


## 14.3   WORKFLOW

No matter how clever a case management system gets, and they can be very clever, they are not workflow systems while they rely on humans to undertake tasks and process cases.  In a true workflow system the workflow program becomes a user in its own right;  either as a constant nag on the case-holders or actually undertaking its own appointed tasks according to rules set down by the organisation.

A workflow system typically not only needs to have access to the kind of information maintained in a case management system but also needs far more data. Rather than simple details about likely "next steps" it needs to be aware of precise details of each decision fork that may be reached in the life of a transaction, together with the rules and criteria for each decision and the full range of tasks which may need to be undertaken in the processing of a case depending on the precise decision route followed.  It also needs to be programmed with all relevant information concerning the "roles" of each of the users together with a table indicating which tasks can be performed by which user "roles".  For example, in a workflow implementation in a law office there may be certain tasks that may be performed only by a partner, others that may be performed by assistant solicitors and still others by legal executives or legal trainees.

The workflow will mimic the standard procedures for a particular type of legal or administrative transaction.  It will determine when the transaction starts, which role type will instigate it and what base data is required.  Thereafter the workflow controls the process according to the rules it has been given.  For example, in the simple, and incomplete, contentious case workflow shown below the workflow system should be capable of sending out the letters automatically, together with the correct names and addresses, as well as knowing which court to apply for the writ and how long to wait before service.

It could be argued, especially with such an example, that the long established debt-collection systems which have been in use for some time are essentially performing the same functions as the one described.  There is certainly a degree of similarity between the two types of systems.  However, the debt-collection systems are usually stand-alone systems beavering away in a universe all of their own with little or no connection with the rest of the office workers or systems whilst the workflow systems, which it should be said are perfectly capable of running debt-collection, are also capable of interacting fully with other users and other systems. Much of the time they spend waiting for users to process tasks so that they can be routed on to the next step.

There is another important difference which should not be underestimated: whereas tasks such as debt-collection and domestic conveyancing are fairly rigid repetitive procedures, almost all other legal matters are incapable of such consistent automation. A Eurobond issue or company merger is a far more complex and varied process which cannot be delegated to a machine in the same manner as debt collection. However, in each of these transactions, no matter how complex, there are a number of sub-tasks which can be largely described in a fashion capable of being assisted by workflow systems. This "cherry-picking" approach, as opposed to outright automation, together with the ability to tweak the workflow in advance and as required for each particular matter will ensure that workflow systems have a place for most types of legal transactions.

The ability of workflows to be chained and to call different workflows in specified circumstances means that a set of basic workflow libraries can be maintained by the firm and be combined on a unique basis for each transaction together with unique elements specific for each case.

Apart from the tasks already mentioned workflow systems are capable, *inter alia*, of the following:

- initiating telephone calls

- scheduling meetings

- obtaining data from internal and external databases

  - updating internal databases

  - running knowledge-based scripts in order to assist in decision making

  - reminding users about due dates and actions

  - sending electronic mail

  - providing automatic document generation.

The first true generic workflow systems appeared in the late 1980s, surprisingly, from this side of the Atlantic. FCMC Ltd, which had developed a very successful financial modelling package, then turned to workflow systems and developed Staffware, a UNIX based workflow system designed to co-exist with UNIPLEX office system. At around the same time a small Dublin-based software house developed Workhorse in association with a Cork law firm. Workhorse was designed to integrate with UNIPLEX and INFORMIX.

FCMC concentrated on selling Staffware to large computer suppliers on an OEM basis, and it was adopted by, among others, Wang and UNISYS. Workhorse Systems sold systems directly, then worked on a Windows client implementation of their system which allowed them to sell it to AT & T as the workflow element in Rhapsody. When AT & T bought NCR, Workhorse was one of the few remnants of Rhapsody to be earmarked for the NCR Cooperation systems, although recently NCR has determined to build its own system.

There are a number of workflow systems which have appeared recently, many having been produced by surprisingly small software houses, such as Softix with their new AXXS system.

Advanced workflow systems permit extremely complex logic which provides the ability to split a workflow into two or more parallel sub-tasks to be carried forward concurrently to rejoin at a later stage, and also allow a sub-task loop to test repeatedly for target criteria to be met before allowing the flow to continue.

Parallel sub-tasks allow different users fulfilling different "roles" to get on with separate tasks at the same time in order to maximise the speed of the process. In an acquisition, for example, the secretary could prepare a boilerplate document while the assistant solicitor gathers the information to be merged with the document and the partner advises by letter on the form of the transaction and the documents to be used.

The sub-task loop would permit a solicitor to fire off a letter to the other side and be reminded weekly to check for a response before continuing in the appropriate fashion upon receipt of an answer.

To summarise workflow systems from the fee earners point of view, what he is provided with on a daily basis is:

- easy direct access to all the information he needs to assist in his current matters

- a list of specific tasks to be undertaken together with all the appropriate tools and information with which to undertake each task. This may be:

- a document to review

- a letter to send

- a draft document to prepare

- authorisation for some action

- a decision to take

- information to provide

- information to request.

Workflows could be established for storing precedents, document preparation and review, retrieving precedents, ordering stationery, individual pay reviews, opening matters and performing automatic conflict of interest checking, closing matters, and so on.

## 14.4   IMAGE

Workflow has long been associated with image manipulation and management, and many of the more recent workflow systems that have been developed began life as image communication tools from image suppliers in order to add value to their image systems. Having developed image file and retrieval systems that allowed users to manipulate image files as easily as other files, the main suppliers then realised that what the users actually wanted to do was pass copies of images between them depending on certain criteria. They then added basic workflow facilities to meet this need, and to provide enough additional functionality to justify the cost of imaging technology.

Traditional image suppliers who have added workflow technology include:

- Philips with ECHO (Electronic Case Handling for Organisations)

- PLEXUS

- IBM Imagesoft.

The still more recent trend appears to be away from file and retrieve systems with limited workflow to functionality which can be used for generic workflow systems, not limited to manipulation and communication of images. In fact, one of the market leaders in this area, Filenet, recently announced that it now considers itself to be primarily a "workflow" supplier rather than an "image" supplier: its new system is known, rather confusingly, as WorkFlo. As an echo of this trend one of the smaller companies, ImageSolve, has also recently announced two new workflow products, ImageSolve II and KeyFlow.

## 14.5   BENEFITS

These different technologies can be implemented singly, or in combination, in order to meet different requirements. Case management can be extremely useful by itself; workflow systems really contain the elements of case management within them and both of these technologies can be enhanced by the addition of image technology. The case management system with integrated image technology provides a method for maintaining the complete matter file on the computer system, including the text of all documents prepared internally, or deemed worth scanning in full text, and the image of all other externally created documents. All these documents are present on the system in chronological or any other order and can be indexed and retrieved on a number of criteria. A number of insurance companies have adopted this approach and banished papers from their offices almost completely.

The addition of workflow technology brings another great leap of functionality. Incoming documents can be scanned in and the workflow can bring them to the attention of all those working on the matter at the same time without waiting for circulation or photocopying. They can also be available to all users for instant recall irrespective of the whereabouts of the physical file. Workflows calling for vital decisions to be taken can produce all relevant documentation for immediate review in order to enhance the quality of decision making that may otherwise get taken by the fee earner in a hurry based on the information readily to hand.

I am under no illusions about the mammoth size of the task in developing even a small number of workflows for the legal office. Many man years of development are involved. However, many of the basic workflow building blocks can be provided by the system providers leaving the individual firms to add value and their own expertise by building their own building blocks, workflow chains and *ad hoc* amendments.

All this is very expensive, both in terms of the technology and the effort required for in-house development and implementation. It may be a while before the technology is cheap enough to warrant large scale adoption. However, there are substantial benefits to be gained by the implementation of such systems and we will review some of them.

### 14.5.1 Procedures

One of the first areas of potential benefit is that of office procedures. These are the things that we all agree are extremely good ideas when we read about them, but which we never have time to put into practice when clients are pressing. We all believe that matters should not be opened without the full and correct information being provided in each instance; that conflict of interest checking should take place when required; that file closure should be undertaken on a timely basis with a review of all documents in order to determine whether they could be maintained as precedents. To be more precise we all believe that other people should be obliged to do these things, but not me, not now, because I am in a hurry.

The inability of law firms to enforce just the sample procedures referred to above effectively can lead to lost know-how, incomplete information, unnecessary re-invention of the wheel and embarrassing and/or costly client conflicts.

Normal methods of attempting to enforce compliance with such procedures are regarded as initiative tests for avoidance on the part of the busy professional. On the other hand, the well constructed workflow system will, in effect, enforce compliance on the part of the fee earner almost unknowingly because it performs most of the task for him on an intelligent basis leaving only a few things to tidy up in the process.

### 14.5.2  Quality

Case management and workflow systems also have a vitally important role to play in the areas of quality assurance and quality control throughout the firm. Once appropriate assisted procedures are in place to capture know-how within the practice and workflows are devised to assist the fee-earner to undertake each transaction in accordance with the firm's considered "best-practice" approach in each case, with the minimum of effort on the part of the fee earner, then quality can be maintained by cost-effective proactive quality assurance techniques as opposed to more costly reactive quality control.

This allows the firm to be sure that matters are being undertaken in the way it thinks they should and that the firms imprimatur is firmly stamped on the service provided and each document produced. Such facilities go a long way towards assisting firms interested in accreditation under BS5750 and ISO9001, but they also go some way beyond that and address other total quality management issues which should, one hopes, result in SIF levy rebates for firms that can show effective implementation of such systems. Research indicates that the vast majority of actions against solicitors is a consequence of simple administrative "cock-ups" such as time limits overrun, rather than incorrect legal advice.

### 14.5.3  Management

Such facilities also provide a powerful means for assisting in the search for higher leverage, or gearing, as they should enable partners to manage and oversee a larger number of matters without prejudicing the quality of the work to clients. This productivity gain would be an important method of increasing the partner remuneration prospects of a firm in an increasingly competitive environment.

### 14.5.4 Group Working

In most law offlces it is not individuals who work on matters, rather, there are teams consisting of partners, fee earners and support staff all working, ideally together, to a common goal. Case management and workflow technologies can enable groups of fee earners and secretaries working together to be far more efficient and effective both as individuals and as a team.

### 14.5.5 Conclusion

All of these factors combined lead one to the conclusion that the time will shortly be ripe for these technologies, combined with other technologies such as tele-communications, to bring about fundamental changes to the way in which "lawyering" is undertaken; in ways which we cannot do more than guess at under the present state of development of these technologies. The quest of any forward-looking law pracffce at the moment will be to seek new ways of allowing it to compete on the basis of quality of service, without unduly increasing costs, and be able to offer increasingly attractive income to partners and prospective partners. There may be many avenues to explore in search of these objectives: case management, workflow and imaging technologies will be at the forefront.

# 15. Support for the Collaborative Working of the Litigation Team

*Morag Macdonald*
*Bird & Bird*

## 15.1   INTRODUCTION

A piece of litigation does not have to be very big or complex before more than one solicitor in the firm becomes involved.  If nothing else, most pieces of litigation will involve the partner and an assistant in any firm.  Larger pieces of litigation can take up the time of yet more solicitors especially during work intensive periods in any piece of litigation such as discovery, production of witness statements and the preparation of the trial and the trial itself.  Also, it can be that one is running a series of actions for one particular client, on which it is important to co-ordinate the various different solicitors who are dealing with those and, of course, report to the client.   These sort of situations require communication amongst the solicitors involved and can be somewhat of a headache for the partner or senior assistant who is managing the piece of litigation or the set of litigation involved.

Information technology has of course helped considerably in this field, but where it is now particularly showing a benefit is in the ability to share information and to enhance communication between teams of lawyers.  What this paper discusses is a few instances of how that is achieved and the benefits and the pitfalls involved.

## 15.2   THE MULTI-LAWYER TEAM

The critical issues in running a multi-lawyer team are, in simple terms, communication, coordination and keeping track of the documents.

Communication is important on all levels and throughout the team.   It is essential that all members of the team know at least what the key issues are and from day to day what new main developments have taken place.  This becomes particularly important at times such as during the discovery process and during the lead up to trial.

*V. Mital (ed.), Advanced Litigation Support & Document Imaging, 159–163.*

The normal way of ensuring proper communication in the team is to have regular meetings and sometimes to circulate copies of correspondence, notes and opinions. However, this can be time consuming and there is often an inevitable time lag for information to be transferred. Also it is often difficult to find times for meetings when the relevant members of the team can all attend, especially when everyone is working to a stiff deadline.

There is no doubt in my mind that a good IT system used properly can improve and facilitate communication in a litigation team enormously. I should explain that my experience in this area is predicated on each lawyer, para-legal and secretary in the team having a terminal or equivalent on their desk linked to the terminals of the other lawyers. This has been the case amongst the litigators in Bird & Bird for some time now. When this is the case the electronic mail system comes into its own. The same message can be sent quickly and easily to several people at once and longer files can be attached to those messages. Queries can be sent to people when you think of them and updates from various members of the team to the overall manager can be copied to others in the team keeping them in touch with what is happening.

Sharing the same area for all files created in-house on any particular action is also an excellent means for coordinating work. A simple example of this is keeping a track of the numbering of party and party correspondence especially just before trial or during interlocutories. I usually find that it is essential to number my party and party correspondence when more than one letter in the day is going to the other side just to make referencing those letters in subsequent correspondence easier. However, different members of the team responsible for different aspects of the case may be preparing different letters to go out in any one day. Provided the computerised workspace is propertly organised it is possible to make this coordination of numbering quite straightforward.

Sharing databases of information can also be invaluable especially when trying to keep a grip on the documents. When someone wants to leave a comment on a document somewhere which can quickly be tied back to the document in question "Post-its" are all very well, but they are easily lost, and it is laborious to search for specific items mentioned. By now I am sure that many in the audience are familiar with the searching and sorting delights of databases. Add to this the ability to leave comments on each document on the database which everyone in the team can see, search on and add to and you have an extremely effective tool.

Sharing of E-mail, documents, spreadsheets, databases and even calendars can enhance the working of a multi-lawyer team incredibly and I believe, lead to an improved work product and service to the client.

## 15.3 USE OF DOCUMENT IMAGE PROCESSING

The benefits of having images of documents available on the screen on your desk are relatively easy to imagine in such a document intensive activity as litigation, especially bearing in mind the premium on document storage space in many parts of the country but, in particular, London. You can always tell when you go into a law firm when you have reached the litigation department because of the amount of paper and files spilling over the shelves and often floor space. What document imaging provides is easy access to your documents and getting rid of bulky dog-eared files.

The document images in question can then be linked to databases like the one mentioned above, allowing for access to the documents by the whole team with easy means of searching for the right document, sorting the documents and commenting on them.

Recently, I was involved in a piece of litigation where there was parallel litigation in the United States. This led to the discovery in the action in the US, which consisted of some 200,000 documents, being discoverable in the English case. The first benefit of having those documents on an imaging system was the sheer bulk of what had to be shipped over to us from the States; a few tape cassettes rather than boxes and boxes of photocopies. Needless to say the job of sifting the relevant documents from this bulk was not an insubstantial task but with shared access and the ability to share notes it was possible for several members of the team to work in parallel on this.

Interestingly, at the initial sifting stage full text retrieval was not particularly useful as each document did have to be looked at systematically. Bearing in mind the added resources overhead in not only scanning all the documents but also putting them through the optical character process, I would question whether it is always advisable to OCR every document which appears on discovery but I think that this varies from case to case.

Also, I do not think that the benefits of image processing are limited to cases with a large number of documents on discovery. However, as yet I am not a complete convert to document imaging systems. None of the ones which I have seen to date completely meet the needs of the litigator in terms, if nothing else, of simplicity of use for everyone in the team or integration with other parts of the lawyer's normal working system. There is also a resources problem in terms of actually feeding the documents into the system, which has to be addressed in each case and which might put a large hidden overhead into the case which is not properly appreciated. Then again there is a capital outlay issue as many of these systems at present might be relatively costly to implement especially across a complete litigation department in terms of hardware, software and training and without the system being available to every member of the litigation team many of the benefits of a system like this just would not be available.

## 15.4   MONITORING ASSOCIATES AND ASSISTANTS

One of the biggest problems for the person managing litigation is keeping a weather eye on what all the relevant assistants are doing.   Once again, however, good computer systems can assist enormously.   Databases can be created for such managers to enable them to keep a track not only of what each assistant should be doing at any given time but also financial and billing details on the file.

Once again, the shared workspace on the system for any given file gives an excellent opportunity for monitoring what is happening on it minute by minute.  I have sometimes picked up draft letters produced by my assistants, which they have not yet put on my desk and sent them an E-mail message approving them or, indeed amending them first, before the assistants in question have actually mentioned them to me.  It has to be said that this normally only happens on files which are moving very quickly but nevertheless it helps to speed them along even more.

Of course, good communication assisted by an E-mail system is an enormous aid in monitoring as is the ability to look at the same document at the same time. Often I have found that when I am in Court in one matter, when I will not get back to the office on a regular basis before my assistants have left for the evening and where I have often left the office in the morning before they get in, I can still monitor and communicate with my assistants on other cases very effectively, simply by leaving them E-mail messages when I get back to my desk and their responding in kind.   Clearly, this will never substitute for a face to face conversation on a complex issue but it certainly aids management in difficult circumstances.

## 15.4   CONTRIBUTION TO LEGAL EFFECTIVENESS

On a recent application to the European Court of Justice one of our European law specialists collaborated with me on putting together the necessary documentation and legal argument.  However, in our firm the European lawyers are on a different floor from the IP litigators and therefore, potentially, we had a communication problem especially with the delicate art of putting together suitable legal argument. Being able to look at each other's drafts and make appropriate amendments on screen saved a considerable amount of time and effort and, I believe, led to our concentrating our effort on the important aspect of dealing with the law rather then how to coordinate our interaciton.

This is probably only one small example of the way in which legal effectiveness can be improved by a good computer system but it may not be one that is immediately obvious. Much more obvious is the possibility for sharing information and, in particular, precedents, opinions and notes. Editing of such things should also be capable of careful control so that precedents, in particular, can be kept in a useful up to date state updated by those who have the necessary specialist knowledge but useable by others.

For example, I have a selection of prayers which can go on an IP Writ which any one of even my very junior assistants can use to produce a decent first draft. It is then usually a matter of minutes to amend such a draft to deal with any issues which they have missed out or which I feel could be worded better.

## 15.5    CONCLUSION

A well set up litigation system can be of enormous assistance to a litigator particularly in a multi-lawyer or multi-party situation. Even a basic E-mail system can aid communication and coordination. However, the real benefits are really only felt when the whole team has reasonable access to the same IT system and the ability to access such a system includes having been given sufficient training to do so.

# 16. Litigation Support for Construction Cases

*Michele C. Gowen*
*Gowen Deets*

## 16.1  INTRODUCTION

The construction industry in the 1990's is one, if not the largest international industry. We encounter the construction industry daily. Whether it be the schools our children learn in, the highways we travel on, the buildings we work in, the hotels we stay in, or the power plants that generate the energy we use, all of these structures are created by an intricate process that must bring together the talent and work of many different disciplines and trades. The construction process involves numbers of people, many critical events, intricate schedules, a multitude of documents, and a great deal of money. Millions and millions of dollars are spent each year by governmental entities and private owners to build new projects or to renovate or rehabilitate existing structures. The process by which buildings, bridges and highways are built is long, complicated, and difficult.

The potential duration of construction projects varies greatly according to the nature of the work involved, and the duration of a particular project will, in itself, affect the way in which the parties will approach the work — and the way in which litigators will approach any subsequent dispute. A foundation subcontractor may have a contract with a general contractor to furnish caissons and pilings for an office building. The duration of this type of contract might be six months to a year. By contrast, to construct a nuclear power plant from beginning to end may take fifteen to twenty years. Renovation and upgrading of an existing newspaper plant typically takes two years. In each example, however, regardless of duration, multiple parties, documents, and events will likely be involved. Having a good working knowledge of how the construction process occurs, and having an understanding of the various people and events and how they interact and how they are dependent, are **critical** for construction lawyers.

Because construction disputes often involved large projects with numerous parties, extensive documentation, and many factual areas in which disputes may

*V. Mital (ed.), Advanced Litigation Support & Document Imaging, 165–178.*
© 1995 *UNICOM Seminars. Printed in the Netherlands.*

arise, the need for expert fact management is significant. A key to successful preparation of the facts in construction disputes is control and careful organization of the facts.

Construction disputes arise because the parties involved in a project may have a disagreement among themselves that needs to be presented in a forum to achieve a resolution, or because a problem that gives rise to an issue to direct or third-party liability has occurred on the project. Being able to select and organize the relevant evidence in a construction dispute and prepare it in a persuasive manner is critical to the resolution of the dispute. Construction lawyers are typically faced in a construction dispute with a huge mass of information, some relevant and much irrelevant, that must be reviewed carefully and orderly in an attempt to locate the best evidence to support the client's position. From the data base of carefully selected information, the lawyer must develop a coherent and persuasive presentation of his client's case and prepare other offensive and defensive alternative presentations as well.

## 16.2  PARTIES

As already mentioned, many parties are involved in the planning, design, procurement, and construction of a project. The intricate relationships among the parties involved in a construction project are governed by a series of contractual relationships. The owner of a project is either a governmental entity or a private company that has decided that there is a need to build a structure such as a highway, school, office building, or hospital. The owner will be responsible for securing financing for the construction project.

The architect is normally hired by the owner to design the project and to provide other professional services as defined in an owner-architect contract. Architects are often hired to provide designs years before actual construction begins.

Engineers may be or are usually hired either by the owner directly or by the architect to provide professional engineering services. Depending upon the type of project involved, the design may require engineers representing any of a wide variety of engineering disciplines, including mechanical, civil, structural, electrical, marine, chemical, and nuclear. As with the architect, the engineer provides professional services defined by contract.

On some projects, an owner will contract for the services of a construction manager. The concept of "construction management" is a broad and flexible one, and the precise duties required of a construction manager on a particular project are defined by the owner and construction manager in their contract. However, the general range of duties that the construction manager may be asked to perform include, but are not limited to, formulation of the project budget, furnishing of construction technology information to the architect, managing contract procurement efforts, and scheduling, coordinating the construction activities. As with the architect and engineer, the construction manager is usually paid on a fee basis, which may be based on a percentage of the overall contract value.

The architect, engineers, and construction managers have a professional duty and responsibility to carry out their designated tasks to design, engineer and manage construction projects with reasonable care, diligence, and skill and in compliance with all applicable standards for prudent construction and safety.

The owner may also contract with the general contractors for the actual performance of the construction work. The general contractor may, in turn, contract with subcontractors and material suppliers to assist in constructing the project in accordance with the project plans and specifications. The general contractor generally pays his subcontractors and material suppliers directly and remains responsible for their work, as if he had done it himself.

Alternatively, the owner may choose a multiple-prime contracting format, in which he contracts directly with what under the general contracting arrangement would be subcontractors. In the multiple-prime contracting arrangement, these contractors are referred to as "prime contractors".

Subcontractors or prime contractors are the contractors who perform specific construction activities, such as a foundation contractor, an electrical contractor, an HVAC contractor, or a landscape contractor. They will, in the general contract format, have a contract with the general contractor, and in the multiple-prime contracting format, have a contract directly with the owner.

A material supply contractor is one who furnishes a particular material to either a general contractor, prime contractors, or subcontractor. Examples of material suppliers would be steel suppliers, curtainwall suppliers, and precast concrete beam suppliers.


## 16.3   CONSTRUCTION LITIGATION DISPUTES

Because of the complex factual issues that arise with construction disputes, construction lawyers are critical in identifying, organizing, preparing, and analyzing the extensive relevant factual information. **In many cases, whether a party wins or loses depends on how effectively facts are developed from documents, depositions, interviews, and site inspections.**

16.4    SCOPE OF WORK

Scope of work disputes generally arise from a dispute over the following:

- interpretation of the contract plans and specifications

- timeliness or untimeliness of notice of claims under the contract

- "unforeseen conditions" at the project site.

All of the above may cause what is known in the construction industry as "extra work". The controversy of a change in scope dispute is usually centered around whether a party should be compensated in either time or money for work performed outside the scope of the original contract. Often, defining the original contract scope is as difficult as understanding the subsequent dispute.

Again, familiarity, organization, and control of the contract documents that govern the project and the project documents that discuss or interpret the contract are essential.


16.5    DELAY, DISRUPTION AND ACCELERATION

The most frequent types of construction disputes are delay, disruption, and acceleration claims. Projects are constructed according to a schedule. It is critical to understand how the project was originally scheduled, what the status of the schedule was at the time of the delay, disruption, or acceleration event, and what the final schedule was (the so-called as-built schedule). A delay in performance, a disruption of work, or accelerated performance by contractors will affect the schedule. The delay, disruption, or acceleration by one entity of the work may or may not have an impact on the work of others. Attempting to determine the cause, liability, and damages associated with the delay or disrupted activities is fundamental to this type of dispute. The construction litigation team must be able to focus the investigation and fact development of a delay and disruption claim by understanding specific events that gave rise to the delayed, disrupted, or accelerated events. By reviewing and organizing the project documentation on the key events, the team develops the client's "story" with respect to the claim.

Causes of delay to a project can include disputes over the scope of the work, bad weather, untimely approval of drawings or material samples by the owner or architect, delay in receiving drawings, and late delivery of equipment and materials, to name a few.

## 16.6   CONSTRUCTION FAILURES AND DEFECTIVE WORK

Essential to resolving a construction failure is having a site investigation of the failed structure performed as soon as possible after the failure. It is extremely important that this investigation takes place quickly, because when the wreckage is cleared away and repairs begin, important evidence of cause of the failure may be disturbed and lost forever.

Resolving construction failures is not only substantively complex, but procedurally complex as well.

In a construction disaster, procedural complexity is often created by the sheer numbers of injured plaintiffs as well as the multiplicity of defendants, including owners, contractors and subcontractors. Efficient organization of the claims is critical both to the preparation of defenses and to the containment of litigation costs.

Other elements of the preparation of a construction failure litigation include determining the availability of insurance (both primary and excess coverage of the claims involved in the dispute) and identifying the types of experts who will be needed to defend the claim.

## 16.7   SURETY OR INTERNATIONAL LETTER OF CREDIT DISPUTES

Major construction projects frequently require the contractors performing the work to give a performance bond and a labor and material bond or an international letter of credit. The bonds or letters of credit are issued by a surety company or bank which, if the contractor defaults, will be liable for the remaining work and the payment of suppliers of labor and material to the project.

Often a contractor may default on not one but multiple projects, and the surety or bank must come in and complete the work on these projects by either financing the defaulted contractor so that it can complete the jobs, or bringing in a new contractor to complete the job.

## 16.8   ALTERNATIVE DISPUTE RESOLUTION

The method for resolving construction disputes for the U.S. construction industry is traditionally usually handled by either litigation or arbitration. Because of the expense and duration of litigation, the construction industry has attempted to find alternative means of resolving disputes. Mini-trials, mediation and arbitration are becoming common forms of alternative dispute resolution. However, it is very important to emphasize that having well organized factual systems is just as important in alternative dispute forums as it is in litigation.

## 16.9    DOCUMENTS AND DUTIES

One of the important aspects of construction disputes is understanding the intricacies of the construction industry.  Distinct differences exist among various types of documents and among the duties of various parties in the construction project.  The construction lawyer must become familiar with these intricacies in order to identify the most important documents in a case and evaluate what evidence is important to the "story" and what is not.

On any project there will be a number of contractual agreements that define the responsibilities and roles of the various parties.  Failure to comply with the terms of a contract can be the basis of a claim, and compliance with the terms of the contract can provide a defence to the claim of an adversary.

## 16.10    DOCUMENTS AND THE CONSTRUCTION PROJECT PROCESS

Documents are an important form of evidence and are critical to a construction case because they tell much of the story of the project and are less subject to the vagaries of human memory than oral testimony.  Contract documents, bonds, bidding and estimating documents, reports, drawings, schedules, correspondence and memoranda are all helpful in developing a chronology and litigation theme for the project.

The documents included as "contract documents" are specifically spelled out in the contract for a particular project.  Most commonly included in the contract documents are the plans, specifications, addenda modifying or supplementing the specifications, general conditions, and special conditions.

The plans are detailed drawings of a project, usually drawn to scale.  The plans are prepared by the owner's architect and contain dimensional information for the contractor.  Specifications ("specs") are used in conjunction with the plans to estimate and construct the project.  Specifications tell the contractor exactly what the contractor must furnish.  The architect usually prepares the specifications as bidding documents for a project, as well as the addenda.  Addenda to the specifications are issued when the designedrchanges the design during the bidding period.

Understanding and analyzing the project insurance documents that govern the contract is also important.

Estimating the job is very important for the bidder-contractor and for the owner.

The contractor must be very sure to make no mistakes in bidding the work, computing labor and material costs, and adding overhead profit, or it may well find itself committed to perform the work at a price that is not sufficient to yield a profit or even to cover costs. In the compilation of costs, the estimate often includes a contingency fee. The contingency fee covers unforeseen items such as wage increases or other conditions, as the owner and contactor may agree.

Bonds or Letters of Credit are often required as part of the bid or defined in the contract documents. A bid bond is often required to be submitted with the bid. It guarantees that the contractor will enter into the contract for the price bid.

Other bonds are included as "contract documents". For example, a performance bond guarantees that the contractor will complete the project for the contract sum. The payment bond guarantees that the contractor will pay his labor and material suppliers.

Contracts are **the all-important documents** in a project and in developing a case for your client's position in a construction dispute. Contracts state the duties of each party. The construction lawyer must understand the contract provisions, including, among other things, the changes clause, notices clause, defective work clauses, and limitation of liability clauses. Understanding the language of the contract, the intentions of the parties at the time the contract was negotiated, and the course of conduct of the parties during the performance of the work is fundamental to developing the litigation theme and strategy.

In addition to contracts, when contracts are awarded, the owner or architect will give the contractor a notice to proceed, after which work on the job can officially begin. The contractors must keep accurate records of the daily "happenings" on the project. The progress report is used for this purpose and its preparation is sometimes a contract requirement. The project manager, job superintendent, or others on the job should keep progress reports daily, weekly, and/or monthly to record what has happened on the project. Another related document that is often a contract requirement is the job conference minutes report. Job conferences are usually held weekly or monthly, depending on what the contract requires. Often, the owner, architect, general contractor, or others present will keep minutes of the meetings for their own records. Usually the architect or general contractor will send proposed minutes to others involved in the project to be reviewed for discrepancies. If any discrepancies are found, they should be put in writing and sent to all those who received copies of the minutes. The job conference minutes are important because they are used for coordination purposes among contractors, subcontractors, and suppliers. The minutes are also a written record of what was discussed, promised, and achieved at the job conference meeting and thus constitute important evidence in any subsequent litigation.

In addition to these documents, the scheduling documents set forth by contract requirement are very important. In a new job a preliminary schedule will determine delivery of material, equipment, and workers to the job. Schedules are an organized way of completing the project in a timely fashion. The most popular type of scheduling is the bar chart, which charts each area of the project from start to finish.

Most projects are designed in their entirety before fabrication and erection. Other projects are "phased". In a phased project, each area of the project is designed and constructed separately.

Many projects follow the "critical path method" of scheduling. This computerized method is used to plan, schedule, coordinate, and evaluate the construction activities showing the amount of time needed to complete each activity. Construction activities are dependent on the completion of the preceding activities.

During construction, unforeseen conditions, owner changes, and other events may change the work defined in the specs. A change order is a written authorization to perform additional work or omit specified work in a contract. A field authorization order is an authorization to perform work pending the approval process of a change order.

When the project is nearing completion, usually the architect will thoroughly inspect the project and compare the completed work with the contract documents and change orders. The architect will then prepare a "punch list", a list of uncompleted or corrective items of work to be performed in order to complete the contract.

## 16.11  PERSONNEL

Just as important as keeping track of the documents is keeping track of the "cast of characters" involved in a project. Usually, the cast includes a project manager within the owner's own organization, whose job is to coordinate information between the contractor(s) and the owner. The architect and each contractor and subcontractor will often have a similar project manager within their own organizations.

Most important in a construction project, and many times in construction disputes as well, are the people who actually perform the work, such as estimators, schedulers, and laborers. The estimator, for example, will translate the information in the plans and specs into quantities and prices that represent the estimated price of performing the work. Often, after construction is already in process, another estimate will be performed to determine what the actual cost of a project will be in order to record differences in the estimated cost and the completion cost.

## 16.12  ORGANIZING THE CONSTRUCTION CASE

As each construction project is different, so too is each construction case.  Because of the distinct differences between cases, there is no one correct method to organize facts in a construction case.  The next part of this paper suggests certain tools that are helpful in organizing and developing facts.

## 16.13  INITIAL INVOLVEMENT

The legal team, client, and expert witnesses should meet to discuss at the outset of a construction dispute.  The construction lawyer should review, early on in a dispute or potential dispute, the contract documents and any other pertinent project documents that the client or legal team identifies.

In addition to reviewing documents and meeting with clients, another step that should be taken at the outset of the problem is for the construction dispute team to visit the project site.  The documents that are read and the facts that are discussed will be far more meaningful to the team if they have seen and walked through the project.  From this initial meeting, site visit, and review of pertinent project documents, the legal and factual framework and strategy of the dispute will emerge.  This initial framework will provide guidance to all concerned for the dispute resolution process.

If the project is still in progress when a dispute occurs, and formal claims have not yet been made, the lawyer should suggest to the client that photographs and perhaps videotapes be made to record and document the construction conditions.

## 16.14  CASE OUTLINE

A subject matter outline should be prepared by the lawyers and client to outline the disputed fact issues.  The case outline is an important element of fact preparation to insure the best fact management for a particular dispute and, more important, to focus the team's attention on the critical issues involved in a complicated situation, so that the fact development and dispute resolution process can be pursued as efficiently and as cost-effectively as possible.  A good case outline will allow construction lawyers and the client to cut through the myriad project documents, events, and personnel to focus directly on the problems.

Subject matter outlines can take various forms and will differ greatly from case to case.  The outline where an actual dispute has been filed should include the following:

- plaintiff's claims
- the facts that support the claims
- the defenses
- the fact issues that support the defenses
- counterclaims
- facts supporting counterclaims
- cross-claims or third-party claims
- facts supporting cross-claims or third-party claims.

The subject matter outline should be as manageable and precise as possible. Adding too much to the subject matter outline will decrease the efficiency and validity of the outline. Another reason the outline is so important is that it will form the basis for establishing your files, reviewing your documents, and designing a method of fact retrieval. Depending upon the type of litigation, consideration should be given to a general or specific subject matter outline.

A general outline is one containing broad subject matter categories. An example of the general approach in a hypothetical construction case would be:

I.  Estimating
    A. Pre-Bid
    B. Post-Bid

The specific outline would be one that attempts to narrow the issues and focus on specific issues. An example of the specific approach is

II. Estimating
    A. Widget Company's failure to estimate properly the following quantities, i.e. concrete, steel and excavation costs.

The determination of which method is best for the outline can only be made after a thorough review of the available documents and the case law involved. Usually, the specific method is better for smaller cases and the general approach is more appropriate for the larger complex document cases. Often it is a futile effort to try to define issues very precisely early in the large case because of the overwhelming number of facts and documents involved; therefore, the general case outline is more effective. As the large case progresses, the outline can be redefined and narrowed.


## 16.15  TIME LINE AND CAST OF CHARACTERS

Two other fact tools that should be considered for use by the construction dispute team are a time line and a "cast of characters" list.

Since construction projects occur over a period of time, determining the events that occur on the project and give rise to the dispute is critical. The construction legal team should prepare a time line that depicts in chronological order the important events of the project. The time line will be very valuable to all concerned when reviewing documents and interviewing witnesses.

Another tool is the development of a "cast of characters" list of the important people involved in the dispute. The list should include not only the personnel from your client who are involved but also the important people associated with the other parties as well.

When preparing this list, the legal assistant should include the person's name, title, and company affiliation.

## 16.16  DOCUMENT INDEXES

A file index recording all documents that the lawyers have received, whether from the client or through another party's production of documents, should be prepared. This list should be circulated to all members of the litigation team.

The method of organization of the construction files is determined initially by the number of documents that are involved. If there are relatively few documents, one should set up indexes in several ways in order to maximize retrieval. Examples of different indexes are as follows:

- chronological Receipt from Client or Other Side
  description of the particular file, who transmitted the document to your office, whether the documents have been produced, and whether they are privileged or relevant.
- chronological Files. Documents organized by year.
- subject Matter Outline Files. Documents arranged by outline categories.
- witness Files. Files organized by witness.

## 16.17  CONSTRUCTION TEAM TASKS AND RESPONSIBILITIES
FOR DOCUMENTS PRODUCED

The following part of this chapter is an outline prepared for both the lawyers and their assistants involved in a construction dispute that has actually reached the formal dispute stage. The tasks involving documents, witness preparation, file organization, and trial or alternative forum responsibilities are similar for litigation arbitration, mini-trials, and mediations.

## 16.18 DOCUMENT PRODUCTION AND ORGANIZATION

I.  Production of Documents
    A. Compliance and/or objections
    B. Evaluation of documents that are responsive
       1. Interview with client
       2. Volume
       3. Privileged and confidentiality reviews.

II. Considerations dependent upon B (1) and (2) above.
    A. Duplicating
    B. Microfilming/fiche
    C. Numbering
       1. Distinct series of numbers for each party's documents
       2. Initial document inventory
       3. Reasons for numbering
          a. Easily cited in briefs, motions, etc.
          b. Document retrieval.

III.  Organization of documents — post formal production
    A. Organize documents by chronological receipt.  Index should include: description of particular file, who transmitted, whether documents have been produced, and whether documents are privileged or relevant
    B. Organize documents in chronological order
    C. Organize documents by subject matter outline codes
    D. Maintenance and use
       1. Carefully plan and choose system to be used
       2. Importance of daily maintenance.

IV.  Set up retrieval systems
    A. Decision whether to use manual or computerized system
       1. Number of documents
       2. Number of complexity of issues
       3. Cost-effectiveness
       4. Discovery schedule


    B. Manual
       1. Examples
          a. index cards or coding sheets
          b. Chronology sheets

        c.      Information to be included: document numbers; type of document (letter, memo, etc.); date; number of pages; author; recipient; carbon copies; blind copies; subject matter; importance; name of coder; date coded; and whether the document contains confidential material, contains handwritten notes or marginalia, is privileged, is responsive to an interrogatory, is in response to a document production request, or is a deposition exhibit.

  2.  Revising and updating
      a.      Flexibility of outline and system as new issues develop
      b.      Controls to insure accuracy of information being coded and efficient retrieval.

C. Computer
  1.  Determination of cost-effectiveness
      a.      Cost of lawyer-paralegal time
      b.      Number of documents, depositions discovery, schedule, etc.
  2.  Choosing vendors
      a.      Know computer terminology, e.g. discovery schedule, complexity of issues
  3.  Decisions
      a.      Whether to code documents full text of by subject matter
      b.      Whether coding will be done in-house by vendor
      c.      Information to be retrieved - see section B(1) (c)
      d.      Prioritize coding of files
      e.      Flexibility of program and services offered by vendor
      f.      Clients should be given choice of using their own hardware-software.

## 16.19 DEVELOPING FACTUAL RECORDS FOR USE IN ANY CONSTRUCTION DISPUTE FORUM

I.  Organize fact files
  A. Retrieve documents, deposition testimony, etc. from either manual or computer system
  B. Create notebooks as described in deposition notebook procedure

II.  Prepare fact memos analyzing subject matter outline
  A. Include sections on discussion of facts, documents reviewed, personnel involved, recommendations (e.g. any depositions should be taken, further document review, further interviews, and whether fact area should be developed further)
  B. Develop charts of milestone events and chronology of milestone events

III.    Working with expert witnesses
    A. Locating expert witnesses
    B. Serving as liaison between litigation team and expert
    C. Assisting in preparation of expert witness testimony.

## 16.20  TIME LINE

| Date | Event |
|---|---|
| January 15, 1981 | Contract No. 345 executed between Company A and Company B. |
| February 10, 1981 | Change Order No.2 Executed to Contract No. 345. |
| February 17, 1981 | Late delivery of steel to project site. |
| March 7, 1981 | Job conference meeting.  All contractors notified of delays caused by late delivery of steel. |
| June 5, 1981 | Project schedule shows three to four weeks delay to project completion. |
| July 1, 1981 | One month time extension given to project completion date. |

## 16.21  IMPORTANT PEOPLE

| Name | Position |
|---|---|
| Joe W. Abrams | Estimator for Company A |
| William C. (Scotty) Bull | Job Superintendent for Company A |
| Jay W. Field | Foreman for Company B |
| Deborah E. Fitzgerald | Contract Administrator for Company A |
| Stephen A. Jones | Owner's Representative |
| Thomas F. Larsen | Project Manager for Company B |
| Sarah W. Pearson | Project Site Secretary for Company B |
| Ralph C. Smith | President and CEO of Company B |
| Paul W. Smith | Project Manager for Company A. |

# 17. Data Processing in Financial Investigation and Litigation Support

*Maurice E.F. Fitzmaurice*

## 17.1    INTRODUCTION

### 17.1.1   Subject Matter

This paper is about the use of computerised data or fact processing with particular (though not absolutely exclusive) reference to its use as a litigation support tool.

In presenting the subject, I shall endeavour to give some answers to three broad questions, namely:

(i)     what is data or fact processing and what are data processing systems

(ii)    what sort of cases give rise to a need for it, and what are its particular uses within those cases

(iii)   how will a data processing system fit into a litigation support operation.

### 17.1.2   Layout of the Paper

Section 2.1 of the paper gives an extremely simplified introduction to the meaning of data and to the way in which the computer organises it, and to the meaning of data processing and the way in which the computer achieves it.   Section 2.2 gives a number of examples of the kind of useful things which data processing can achieve.   These examples are in the abstract - i.e. they are not linked to any particular case background.   Section 3 provides actual case examples of the use of a number of the data processing techniques described in Section 2.2.

While this order, from theoretical background, through abstract examples to actual case examples, is the logical one, some readers may prefer to start straight

*V. Mital (ed.), Advanced Litigation Support & Document Imaging, 179–209.*
© *1995 UNICOM Seminars. Printed in the Netherlands.*

away with the case examples and work back to the more theoretical aspects of the subject thereafter.

The final Section 4 covers briefly a number of topics which are relevant to the question of the use of a data processing system in the context of litigation support.

### 17.1.3    Document Handling/Data Processing

I shall be describing the nature and characteristics of data processing in some detail below but I shall give an initial idea of this right at the outset by making a comparison between data processing and text handling, which is the subject matter of the majority of litigation support systems with which you will be familiar.  I shall do this by postulating a very simple distinction between the function of a data processing system and that of a text handling system.  I have no doubt that like many simplifications, this distinction is not universally true or accurate.   But I believe it is both broadly true and also illuminating of the underlying nature of the two forms of systems.

In the overall operations typically using a document handling system (a system which comprises the storage, indexing and retrieval of texts), the major processor involved is the human, and more specifically the lawyer's, brain.  And what this human processor is doing is thinking.   That is, reasoning, developing arguments, analysing and presenting issues and so on;   typically, things the human brain does well but which the computer still does virtually not at all.   The role of the computer in these operations is to enable the user to get at the information contained in texts of any kind which he needs to support the processes of his thought.   It is, as it were, largely confined to the role of a fetcher and carrier of information.   It does not itself to any significant degree process the information concerned itself.    A simplified diagrammatic representation of this appears in **Figure 1**.

Notwithstanding this, much of the information to which we may gain access using a document retrieval system will require processing in some way before we can make use of it.    Some of this processing will be entirely beyond the capacities of computers.    For instance, we may want to make a summary of the contents of a document, either to store for future reference or to incorporate into argument, or a pleading.    The production of such a summary is totally outside the range of capabilities of a computer.    But some of the processing we may want to have done will be potentially within the range of a the computers capabilities; and may, indeed, fall within areas where the computer is particularly efficient and the human brain particularly inefficient.

For instance, in a complex construction case, we may want to have a summary of actions taken, or not taken, in the course of a particular stage in some part of the contract works; and we may want to know how those actions, or their timing, was affected by some actual event, or would have been effected by some hypothetical event.

In a case involving insurance and re-insurance, we may want to have a summary of the re-insurance risk of individual underwriters, or shareholders in re-insuring companies at any particular time, or at any particular place in the re-insurance chain and compare it to their risk under the main insurance policy.

In a case involving money laundering, or the manipulation of accounts, we may want to view the transactions on a complex web of bank accounts in many different ways; for instance we may want to see the results of consolidating the accounts together, or we may want to view the comparative movement in the balances on accounts over a period; or quantify the flows of money between the accounts over a period.

Any of these things could be done entirely manually, or by using a document retrieval system to get at the documents in which the information necessary to perform the processing is contained and then using our own processing power to obtain the results. But either of these methods would, unless the number of facts or transactions involved was very small, be extremely inefficient, time consuming and costly as we are not really at all good at assembling masses of individual items of data, performing calculations on them, recording the results, and so on, in a repetitious way.

By contrast, computers are absolutely brilliant at doing these things, provided that the data involved can be organised in a way in which the computer can make sense of it. Happily, there are such ways of organising data; and it is in consequence possible to provide a tool for, among other things, litigation support which will perform the greater part of this type of processing by computer. Such a system is what I am calling, for the purposes of this paper, a data processing system.

**Figure 2** gives a diagrammatic representation of the use of such a system, which may be contrasted with the similar representation of a text handling system contained in **Figure 1**.

The ways in which the information concerned must be organised if the computer is to be able to process it, and the limitations which this imposes on the kinds of process it can perform, and the kinds of case in which it may be useful is covered in detail below.

### 17.1.4   Development of MFIND

It may be observed from the brief autobiographical details provided somewhere in the conference documentation that my relevant experience which started in the field of law, as a commercial barrister in London, passed through a number of years of banking and financial investigation, and ended up with the development of a computer software system, known as MFIND, to assist in financial investigation and general data processing.

Most of the data analysis techniques that I shall be referring to in this paper have been incorporated into MFIND; and I shall be taking a number of the examples of the use of data processing in relation to litigation from cases in which I have used MFIND.   But the concepts and processes which I describe are nevertheless largely generic and capable of incorporation in other systems.

## 17.2   DATA PROCESSING

### 17.2.1   Definition of Data Processing

#### *17.2.1.1 Terminology*

We have become accustomed to distinguish, in a general and non-technical way, between a number of different computer systems which are widely available on the market.   In particular, one can mention:

- wordprocessors
- spreadsheets
- data bases
- desktop publishing systems
- project management systems
- litigation support or text retrieval systems

Each of these is aimed principally at the performance of a particular set of functions (though some of these may overlap).  All of them involve the storage of information, and in the wider sense they all, therefore, may be said to incorporate a data base.

It may be seen from this that it is the functions which really distinguishes the various forms of computer system; and to define what we mean by a data processing system, we will have to look principally at what that system is to do, at what its function is.

Before going on to do this, however, we must note that there is a problem of terminology which arises from the looseness with which we are inclined to use terms in relation to computer systems.  Thus, in the wider sense, any information may be properly regarded as data.  For instance, texts, or the contents of texts, or any other form of information, may quite properly be referred to as data and any computer system which stores information may be referred to as a database. In this wider sense, for instance, a text retrieval system incorporates a database in which the units of data involved are the texts, or sections of text, stored in the system; and, indeed, the term database is often (and not incorrectly) used to describe the storage element of a text retrieval system.

However, in almost universal general use, the terms "data", "database" and "database management system" have become attached to systems which, at least by comparison to other systems, are of a generally similar nature to the systems I shall be discussing in this paper. This usage is so widespread that the introduction of other terms, such as fact base, or fact analysis, which I might prefer, may seem pretentious and be confusing.

I am therefore going to use the term "data" to describe the kind of information which is the subject matter of the systems and the term "database" to describe the storage element of those systems.

### 17.2.1.2 What is Data

We have referred above to the simple distinction between text handling and data processing systems. In a text handling system, the typical units of information are texts or sections of text, the significance of which is not machine readable. It is retrieved to be looked at and used, along with other sections of text, and interpreted by a human agent.

Data processing systems are aimed, by contrast, at the performance of necessary tasks on information where computer processing is superior to human thought processing.

The computer's superiority in these respects may be categorised as follows:

- a capacity to locate with almost instantaneous speed, and perfect precision, an enormous (in practical terms almost limitless) number of individual locations within its own storage system

- a capacity to perform a number of different operations on the data stored at these locations, in particular to perform multiple complex calculations on numeric data and to compare the contents of different locations

- an important development of the second of these capacities is the ability of the computer to be programmed to recognise and compare patterns, and in particular, by comparison to the very limited capacity of humans in this respect, to do so where the number of potential variables in the pattern is large.

On the other hand, a computer's capacity to attribute significance to an individual piece of data in its storage locations looked at alone is very limited. We make something of the significance of words by attributing meanings to them and by associating a number of them together in a complex system called a human language. A computer is very bad at attributing significance in this way; and we have to find other ways of attributing significance to a word or numeric figure if a computer is to be able to do anything useful with it.

One of the ways in which we can organise information in a way that allows a computer to attribute significance to it, and probably the most important in relation to data processing, takes advantage of the computer's capacity to locate locations within its storage system to which we referred above. To enable the computer to do this, we create what we may see as a structure of locations in which individual locations are given a significance. The result is that the presence of information in a particular location in the storage system in itself defines the significance of that information.

But what sort of information must we put in the locations? The simple answer in theory to this is that we should put information into separate locations which has been broken down into the smallest individual units to which it can be reduced. Looked from a more practical standpoint, we may rather say that we should put information into the separate locations in the smallest units in which it will be desirable for the computer to be able to recognise them.

This is best illustrated by an extremely simple example. We may well regard the information contained in a sentence such as "A paid B USD 100 on 01.01.92" as a single piece of information. We are able to do this, however, only because we know a number of things which, from the text alone, the computer would not know. In particular, we know that A and B are people and that A is the payer and B the payee. We know that USD is a money unit called a currency and we know that 01.01.92 represents a particular date.

In order, however, to allow the computer to, as it were, "know" these things, we will have to break the piece of information into smaller pieces (which I will be referring to in the rest of the paper as "units of data", or collectively as "data"). Thus, in the computer's storage system, we would create a little series of locations which would be identified as locations for, respectively, the payor, the payee, a currency, an amount, a date (see **Figure 3.1**). Insertion of the relevant units of data into these locations would enable the computer to recognise what each of those units was, and to perform routines based on the overall information contained in the sentence set out above.

*17.2.1.3 Database Structures*

Taking the process of our series of locations further, it is likely, though not absolutely invariable, that, in any case in which we are going to want to use computer processing in relation to the payment of money, we are going to be dealing with not one, but many payments, probably involving different people, currencies, amounts and dates.  In other words, we are going to be dealing with not only "multi data" (that is, information which can be broken into multiple units of data as described above) but also "multi transaction" cases.

To allow for this, we may adopt a fairly standard system of organisation of the computer's storage system in which we will create an identical little series of locations for each payment, and we would insert the units of data concerning each payment into the locations in the same order.  That is, say, location 1 would have the payer, location 2 the payee, location 3 the currency and so on.

There are a number of terms used to refer to these locations and series of locations.  The most common, which I will use when necessary in this paper, is to call the series as a whole a "record" and the locations in it "fields".  Thus, there will be a record for each payment, which is divided into the same number of fields, into which the individual units of data will be placed.  In order to distinguish the set of records in the system relating to payments of money from other records, they will be grouped in some way within the storage system.  Such groupings are commonly referred to as files.  In a good data processing system, moreover, the computer must be able to locate any particular record (or payment) from the whole group more or less instantaneously and for this purpose, the record will have to be given some form of identifier which will point to its position in the storage system.  I will refer to this identifier as a "key".  (The organisation of just 3 Money Transfer records into a file is illustrated in **Figure 3.2**).  On this basis, the computer can instantaneously locate any particular location in the system, and "know" the significance of the contents of that location, when it is given the file, key and sequential position ("field number") of the particular unit of data.

At its simplest, on this basis, the structure of storage of units of data in a data processing system may be seen as a simple pattern of locations formed into rows (the records) and columns (the fields within the records), the records being grouped, in accordance with the type of information they contain, into files.

The structure of the storage in a sophisticated data processing system will probably be much more complex.  For instance, the fields may be divided and sub-divided and the divisions of a group of fields may be associated in a structured way.

Furthermore, the computer may recognise other forms of link or structured relationship between units of data besides those that arise from the simple file, record, field position relationship. Extremely important additional relationships of this kind are set up using a system which categorises units of data into what I am going to refer to as primary and secondary units. In this system, primary units are units of data which are unique throughout the system. Each such unit is defined, and information about it contained, in a separate record unique to it in a particular file. The easiest example of a primary unit of data is the name of a particular person (which in practice is likely to be a special short version of the person's name which is used to identify that person, and only that person, throughout the data base). There will be a file for information concerning people who feature in the information in the data base in which the record key will be the computer's name for that person. The fields in each record will contain units of information about the person, such as addresses, family connections, company connections and any other information which may be relevant.

In other files, certain fields will require entry of the name of a person with respect to whom there is a record in this file of information on people. For instance, fields 1 and 2 of our example record for a money transfer file will require such entry. As a result, the structure of the database is extended from just the groupings of records and fields within the money transfer file to a link from each payor or payee to that person's record in the people file, and in that record to the various units of information in the fields contained in it. In **Figure 4**, we show this link taking record 1 of the Money Transfer file illustrated in **Figure 3** and showing the links from the payor A and payee B to their respective records in a People file.

The structures to which I have referred so far are structures which are built into the database at the outset. In order fully to understand the processes by which the computer is able to locate locations and data, I have to mention also the records of relationships which the system will build up during input in the form of indexes and cross reference files.

At its very simplest, this process might include, for instance, the production of a simple index to the money transfer file, in which, under the heading for each payer or payee appearing in the file, is stored a list of the keys to the records in which that person appears. Again taking the three Money Transfer records from **Figure 3**, we show in **Figure 5** the way in which the "A", "B", "C" and "D" records in a Money Transfer file Index refer, respectively, to the Money Transfer file record keys in which the parties appear.

The use of such indexes is, in a sense, the complement to the use of the structural link system. The structural system gives meaning to and links units of data by reason of their location. Indexing systems link locations by reason of their contents.

The ways in which data structures may be built, and links set up and recorded is an enormous one which cannot be followed further here. It is intended only that the description I have given should provide an idea of the enormous and sophisticated possibilities which exist, on the basis of which data processing can be taken so much further than the rather standard retrieval and reporting systems that form the repertoire of the majority of so called database systems.

*17.2.1.4 What is Data Processing*

The techniques of data processing form a major subject in their own right which is beyond the scope of this paper. I shall describe here only one or two important techniques which are basic to the whole concept and which may open the mind somewhat to the extraordinary power which may be harnessed to the process of organisation and analysis of data once a database has been set up.

The first of these underlying techniques has really already been referred to in relation to the structure of a database. It is based on capacity which the computer has to find, instantly, locations within the structure of the database. This capacity may be harnessed by programming to perform an almost limitless number of tasks, which, when combined together, make the most elaborate analyses possible.

At the simplest, a routine might involve just extracting the units of data from all of, for instance, the locations which we have referred to above as field 1 of a money transfer file (that is the payers) and making a list of them. The complementary operation based on an index would involve giving the routine a particular name and programming it to look up that name in the index to the money transfer file and then producing a list of all the records in which that name appeared as a payee.

But routines based on this kind of capacity can be enormously elaborated and refined. Thus, for instance, the list of names produced by the first instruction could be refined by looking up all the names in the file of people and including only those who met certain criteria (for instance that they were directors of a particular company). Other refinements could be introduced by reference to other files. For instance, there might be files of other types of transactions (for instance the shipment of goods) in which the parties might feature. Such a file would also have an index, so that each name appearing in one file could be looked up in the index to the other file, and the names included in, or rejected from, the list in accordance with whether any of the transactions in the other file met certain specified criteria.

Furthermore, of course, the system could be instructed, having made such a list of people, or of records of transactions, to present that list, or data based on that list in some specified way; and that specified way might involve presentation of the raw data, or of the data formatted in some way, or of the data after some form of calculation has been performed on it (for instance, the sum of all the amounts included in the final, refined, list of transfers).

These systems have involved what may in a way be regarded as vertical searches, producing in one case a list of people and in the other a list of records having a similar characteristic.    Equally important in data processing, but with perhaps particular reference to investigative uses, is the capacity of the system to be instructed to move from record to record as it were horizontally.  For instance, the system may be instructed to make a list of all the payees of transfers from a particular payer, and thereafter to locate all the transfers of money in which that payee is himself a payer and then to locate the payees of all the transfers from him and so on, forming lists of records which link up horizontally, rather than vertically.

This capacity forms the basis of link analysis systems which are described in a little more detail below.   These systems are fundamental to the use of computers in the tracing of flows of money, as well as in some forms of critical path analysis used in relation to project management.

### 17.2.2  Theoretical Examples of the Use of Data Processing Techniques

*17.2.2.1 Specialised Organisation of Data*

We may also want to have underlying data organised or grouped in some additional, special ways, and the system can be arranged so as to set up the necessary locations and relationships for such organisation during the course of input.

An important instance of this is the system used in MFIND to organise financial or comparable) data into accounts.   The underlying data concerned consists of transfers of money, which are organised into a standard structure of fields and records in a money transfer file in the way described above.   The system involves setting up a series of additional locations and cross references which enables routines to read the underlying data as a series of accounts, and to manipulate and analyse them.

To appreciate what this system can do in practical terms, it may be looked at from three angles.

(i)    It operates where the data involved is already organised in a system of accounts; for instance, the normal organisation of a series of transfers of money through bank accounts.   In this case, the main feature of the system is its recognition of the fact that a bank account is actually only a grouping of a series of individual transfers of money. In particular, at the input stage, we are only concerned with input of each transfer as a separate item.   We leave it to the computer to effect the organisation of these individual transfers into accounts.

(ii)     Having given the computer this capacity gives us an extremely powerful tool, as we can also organise transfers of money which are not actually grouped together as an account, or of which we may never see an account statement, into an account. For instance, many complex financial transactions involving banks do not give rise to an account as such at all.   An example is the placement of funds on money market deposit.   An "account" organisation system allows the grouping of individual transactions of this kind into accounts which make them more readily understandable, and allows them to be compared and consolidated with other groups of transfers, such as those passing through current accounts.

(iii)    While mention of a system of accounts brings immediately to mind transactions specifically involving the transfer and storage of money, in fact many other forms of transactions are susceptible to organisation in an identical manner.   Such transactions have characteristics which may be referred to as quantifiable directional flows.   In relation to a transfer of money, it is a flow of value from one person (or account) to another which is quantifiable in units of money.   The shipment of goods is equally a flow, this time of the goods concerned, which may be quantifiable either in money value or in some other system of quantity, such as weight.

*17.2.2.2 Consequential Transactions*

Another way in which the computer can process suitably organised data is by itself processing data from what may be referred to as a primary transaction into data comprising a secondary transaction.

For instance, we may have details of a series of dealings on the foreign exchange market (e.g. spot trades by a bank on behalf of its customer).   Each of these trades will give rise to a series of money transfers in the currencies concerned, both as between the bank and its customer and as between the bank and its counterparty in the market.   The computer can set up accounts and record the postings to them which would have arisen on such accounts (see below 17.3.2.4 for a brief description of a case in which such techniques were used).

*17.2.2.3 Creation of Summaries*

The creation of summaries of data is of enormous importance in both the development of arguments in a case and the presentation of evidence at a trial. For instance, in the case of an account made up of a series of individual transfers of money, a full statement of the account may be of little value in either the development of our thought, or in backing up the argument in court.   What we are likely to need is to look at the individual transfers on the account summed up and/or selected in some form.

We may want to see them in the form of gross credit, gross debit and net movement within a series of periods; or we may want to see them analysed in accordance with the counter parties or accounts involved. We may want to see the balances on one account compared over a period with the balances on other accounts, or to see the details of a group of accounts consolidated as if they formed a single account.

Producing such summaries manually is an inordinately time consuming process. For the computer operating on a well organised data base, it is a matter often of no more than a few minutes.

### 17.2.2.4 Link and Flow Analysis

A particularly important and specialised form of the data selection process referred to in 1 above is what is generally called link analysis. As has been stated above, one of the aspects of the system of structured locations within a computer's data storage system is that the computer can recognise links between units of data. This may be because of the relative position of the units within the record of an individual transaction of a particular type; or it may be because the units occupy similar positions within similar records of other transactions of the same kind; or it may be through the existence of a special record for the unit of data concerned (a primary unit - see above 2.1.3) existing somewhere in the storage structure, in which other units of data relevant to the primary unit of data are stored.

As a result of the combination of all these links, the units of data in a data storage system may well form a complex web of varied links. The simple beginnings of such a web are illustrated in **Figure 6**. Here the starting point is the Money Transfer Index file (see above 2.1.3 and **Figure 5**) for the person "A". From there the first link is to the records (keys 1 and 2) in which A features as payer. The next link is from the payer to the payee, in the case of record 2 called "C". The next links go from C via the record for C in the People File to C's residence and to a company in which C holds shares. Each of these data units is a prime unit (see 2.1.3 and **Figure 4**) and therefore links to its own record in, respectively a Property File and a Company File, where it is linked to yet further data.

The ways in which units are linked to each other directly, and the patterns of links between different units are matters which may be of the greatest importance, both in investigation as well as in financial analysis and in the development of argument and presentation of data in evidence.

Search for, and recording and analysis of, such patterns of links is time consuming to the extent of almost total impracticality by manual methods; but falls fully into the areas in which computers are particularly adept.

*17.2.2.5 Specialised Reports*

Another form of analysis technique available within a data analysis system is the regular production of specialised reports which draw on the data as it may be from time to time during the development of the case.

Such reports can, of course, include a series of formats of statements and summaries of the accounts into which transfers of money (or other transactions to which the account organisation system is applicable - see above 2.1.1) are organised. Because all the systems producing such statements and summaries operate directly on the underlying raw data, they require no amendment in themselves to reflect amendment of data or fresh input, all of which will automatically be incorporated into the statements when they are printed out.

But there are numerous other forms of report which may be of great use. Examples of these are chronologies of time related data (i.e. chronological reports of everything which is recorded in the database as having happened on a particular date, which may range from the opening of a bank account, formation of a company, transfer of money etc. to appointment of a party as a director of a company, purchase of a property, issue of an insurance policy).

Another form of report which may usefully be produced direct from a properly structured system would be a profile of a particular party, or of parties in a particular category, recording in summary form all the information in the data base about them.

*17.2.2.6 Specialised Presentations*

Another form of processing of which a fact analysis system is capable is the production of varied forms of representation of the same data. In its simplest sense, this may be taken to include the various formats in which, for instance, a statement of financial data might be printed; but it includes also the possibility of production of data as well in forms such as flow charts and organigrams, and the graphic representation of data in graphs.

## 17.3 CASE EXAMPLES OF USE OF DATA PROCESSING SYSTEM

### 17.3.1 Reconstruction of Trade Transactions from Incomplete Data

We used the simple data selection and comparison capacities of the data processing system in a case in which a company had attempted (fraudulently) to superimpose documentation which apparently justified payments of money into its accounts on to a series of existing deals under which they had no entitlement to payments of any kind.    The underlying complete deals all concerned imports of raw material.

We had a very substantial body of documents relating to the real transactions, as well as the fraudulent documents produced by the company;    but these were in no particular order or grouping and were incomplete.    They included correspondence between the importer and the suppliers, correspondence between the importer and their bankers concerning the opening of letters of credit;    banking documents relating to the opening of letters of credit;    pro-formas and final invoices; certificates of origin; certificates of insurance;    and shipping documents such as bills of lading and packing lists relating to the materials.

What we needed to do was to find among the documents sets which could be shown to go together and to evidence the existence of a complete (and completed) deal, independently of the alleged intervention of the company as allegedly proved by the fraudulent documentation.    The problem was that there were no transactions with respect to which there was a complete set of all the genuine documents;    and there was no single item of data (for instance a particular reference number) which would necessarily appear in all of the documents in a set and from the presence of which the relationship of each document to a particular transaction could readily have been inferred.    The problems arising from this were added to by the fact that an employee of the company (who was also involved in the fraudulent super-imposition of the false documentation) had deliberately made multiple orders of identical quantities of the same goods at the same prices for shipment on the same date and vessel.

There were, however, a substantial number of pieces of data which appeared in the documents which could assist in stringing documents together to form identifiable individual transactions.    These included such things as the values, quantities (gross and net), packaging details (quantities of bags etc.), dates, vessels, invoice numbers, bill of lading numbers, letter of credit numbers, customer references, packing numbers and so on.

While, as referred to above, no one of these identifiers appeared on all the documents necessary to prove a completed transaction, many of them appeared on more than one such document, and many of the documents had more than one of the identifiers.    We made a data base of all these identifiers and were able to use the data search and matching capacities of the computer to search for and analyse strings of documents which could be shown to have belonged to a single and completed transaction.

The basic way in which it was possible to string together sets of the genuine documents is illustrated in **Figure 7**.

Of course, this job could, like most of the jobs done by a computer, have been done manually.    But the enormous number of transactions, documents and identifiers and the almost exponential growth of the possible combinations of them was such that it would have been an excessively time consuming and costly exercise.

In fact, as in our experience is quite often the case, it was only our knowledge of the ability of the computer to perform the task in a relatively painless manner that gave rise to the idea to attempt it at all.    Before they became aware of this, the lawyers concerned in the case had concluded that the state, quality, quantity and incompleteness of the documents made it impracticable to attempt to prove anything from them.

### 17.3.2   Transfer of Money

*17.3.2.1 Reconstruction of Detailed History of Transactions*

Many cases, both for investigative purposes and for the purposes of providing the basis of claims and defences in litigation call for the reconstruction of the movements of funds across bank accounts and of their relationship to the transactions from which they derive.    Such reconstruction, even on a relatively modest scale, is a daunting task.    On a large scale, something which is not at all uncommon, it was the view of major accountants involved in one case on which I worked, that such reconstruction was impracticable on the basis of methods open to them to the point of being effectively impossible.

That case involved an investigation and analysis of the turnover on a group of over 100 accounts with some 12 banks over a period of 5 years.    The accounts included numerous different forms of account, including current accounts denominated in a number of different currencies, as well as accounts built up at the input stage from multiple placements on money market deposit and many special accounts such as accounts arising from post import financing.

The combination of a data base of the transfers of money involved with the account organisation and analysis system which data processing techniques had allowed us to create in MFIND, rendered the task both possible and practicable. The ways in which it did this are numerous and complex, and could well form the subject of a lengthy paper on their own.    I will give only two examples.

(i)  Postings to, and Analysis of, Temporary Accounts

Because documents became available only in waves and were often incomplete, at least at the time of initial input (for instance bank statements initially unaccompanied by vouchers), we were faced with numerous transfers to or from accounts of which we did not at first know the source or, as the case might be, the destination.

The input routines, coupled with aspects of the general account organisation system, allowed the posting of the unknown ends of such transfers to a number of different forms of temporary or suspense accounts.   We were able to devise data processing routines which, using all the information available concerning the transfers, was able to compare the debits and credits to these various temporary accounts, both with each other and with other transfers, so that, with various levels of probability based on criteria input into the system,  we could identify the actual account from or to which the payment was made and the posting to the temporary account could be reversed and made to the actual account.

This function was of particular importance in relation to the identification of payments which were internal to the circle of accounts, so that these would be properly eliminated in the consolidation process (which we will describe below).

In general, using these techniques, the number of significant payments, particularly payments internal to the circle of accounts, which remained hanging lose at one end were reduced to an extent which no longer threatened the validity of the general conclusions we had to draw from the reconstruction.

(ii)  Problems Arising from Missing Records

Notwithstanding the very substantial documentation which eventually came into the hands of the client, there remained significant areas where we lacked banking records for important periods relating to movement on accounts which we knew to have existed.

We were able to fill in many gaps using the database and data system techniques.

For instance, we had documentation concerning many hundreds of Letters of Credit, and the even more numerous shipments under them, the relevant details of which were input to files in the database.   We further had evidence that certain of the shipments, having a particular series of characteristics, evidenced by this documentation had given rise to actual transfers of money on one or other of the accounts in the circle.   Indeed, the implication was that the documentation in these cases had been set up specifically to give a background to the money transfers.

The process of identifying the particular shipments concerned and of translating the data relating to them into transfers of money (see consequential transactions above at 2.2.2) within the computer was relatively simple.   But even where we knew the bank and/or account within the circle, it was still impossible to simply incorporate these consequential payments into the relevant accounts.   This was because the periods covered by the shipments to a considerable extent coincided with periods also covered by records for the accounts concerned, from which the consequential payments had already been input to the money transfer file.   We therefore had to take some hundreds of payments already in the money transfer file and compare them with many hundreds of shipments, so as to ensure that we eliminated duplication. This process involved taking a number of parameters from the data records of the transfers which had been input from the banking records (such as date, parties, if known, amount, currency, reference numbers, etc.) and a number of parameters from the shipment and letter of credit files (both similar parameters to those taken from the banking records as well as additional matters such as the number of shipments made under a particular letter of credit and the total value of the shipments and/or of the letter of credit) and running comparison and matching routines on these so as to identify which of the shipments were already covered by payments, and which were not.

The number of possible parameters to be married up was considerable, though in any given case, particularly in relation to the data from banking records, the number of those parameters actually, as it were, filled in in relation to any single payment, was often sparse. The process involved thousands of comparisons, and the following of many "what/if" type lines. All this would have been so daunting that it is unlikely we would have attempted it manually. It was entirely practicable using data system techniques and the result in fact filled in the greater part of the gaps in banking data.

The result provided a very accurate and well founded view of total turnover on the circle of accounts as well as the basis for an important analysis of the source and destination of the funds which passed through that circle, even during the periods for which banking records were incomplete.

*17.3.2.2 Analysis of Turnover on a Circle of Accounts*

One of the issues in the case to which I have referred above required us to establish the gross turnover on the circle of accounts, as well as to provide an analysis of that turnover in terms of the counter parties from and to whom payments were made into and out of the circle.

Following the reconstruction of the individual transfers on all of the accounts involved in all of the banks, we used data processing routines, some being part of the special account organisation system of MFIND, others involving use of more general data processing techniques, to produce the analyses of turnover which were required. We will describe very briefly two of these techniques.

(i) Consolidation of Accounts

Consolidation of bank accounts in a circle is extremely simple in conceptual terms, involving simply the elimination of transfers of money between the accounts in the circle to be consolidated and the subsequent treatment of all the remaining transfers to or from those accounts as if they were transfers to or from a single account.

It provides an extraordinarily useful and powerful tool in the reconstruction, investigation and analysis of multiple accounts, and the ease with which consolidations can be made and analysed is the criteria of a good system.

In the case we are considering, consolidation was in fact relevant in a number of ways besides the question of gross turnover. For instance, it was an essential element in the process of piecing together and displaying the customer's overall deposit position with each bank from time to time (which position was actually made up at any one time of a large number of individual placements of money on money market deposit, often in four or five different currencies) and comparing this with its overall position on current account.

In order to provide for all the various requirements of the case, we undertook a series of consolidations in gradually widening circles, starting with the consolidation within each bank of various groups of accounts, then proceeding to an overall consolidation for each bank, and ultimately leading to the production of a gross consolidated account for the whole circle of accounts. In simplified form, the concept of these widening consolidations is represented in **Figure 8**.

This gross consolidation involved many thousands of transfers, and we regarded the resulting print out, which covered some 200 pages, as having largely reference or archival value. But within the computer, the consolidation formed the essential raw material for the various analyses which we made.

(ii)  Analysis of the Consolidated Account by Counterparty

A number of issues in the case required an analysis of the parties or types of party from or to which payments into and out of the circle had been made. In fact there were many hundreds of parties concerned, and it was essential to group these in relevant ways in order to be able to reach useful conclusions from the analysis.

We were able to speed the process of making these groupings by use of the general search and matching techniques of the data processing system. We were able to group the parties largely on the basis of data contained in the overall data base. For instance, one group of parties had to include all of those who, during a particular period, had been the issuers of letters of credit, or shippers of goods, originating in a particular country (note that, for the purposes of this particular case, there were Letter of Credit and Shipment of Goods Files as well as a Money Transfer File). All the data necessary to produce a list of such parties was available in the files of data on letters of credit and on shipments.

Other groups were defined by such things as family or company relationships, data on which was to be found in the People File; and again, general data processing techniques allowed us to build up the lists of such groups of parties within the computer.

In the end, we grouped the parties into about 10 categories, leaving a mass of unimportant parties in an "Unselected" category. Using the account analysis capacity of MFIND, we could then analyse the gross turnover on the consolidated account. For this purpose, we divided the overall period of some 5 years into 4 periods. The computers summary analysis of the movements in each category within one such period is reproduced in **Figure 9**.

This summary revealed an extremely significant pattern, namely, that while the major inflow of funds came from the so called Consignee category of parties (the issuers of letters of credit/consignees of goods), the major outflow was to the so called Family Group and Family categories (namely the fraudsters and their group of companies), with only a small proportion of funds flowing out to identifiable suppliers of the goods allegedly shipped.

Obviously, further investigation of some of the figures in the summary was called for. For instance, without further work, the computer could be called on to produce a more detailed analysis of the payments into the consolidated account by the Consignee category (illustrated in **Figure 10**). This revealed that by far the greater part of these payments came from a party NM.AEA. The computer could then be asked to produce a detailed statement of that parties account with the Consolidated account (illustrated in **Figure 11**).

Cross reference from the field "Bank Reference Number" in the Money Transfer File records for the payments concerned led us to a number of letters of credit allegedly covering the shipment to NM.AEA of large consignments of a particular commodity. This led to substantial further investigation, all using particular aspects of the MFIND financial analysis system. In particular, we were able to gain a preliminary view of the general nature of the transactions concerned by use of the Link Analysis system (for a more detailed description of this see below 3.2.3 and **Figure 12**). Following this, the question of how the money concerned had passed through the circle of accounts, and exactly where it had ended up became a major issue. For this purpose we were able to produce a Flow Analysis of the kind also described under 3.2.3 below. This Flow is too large and complex for description in this paper. It revealed, among other things, that one amount of $10,000,000, though initially passed out of the circle of consolidated accounts, was, by a series of complex movements (themselves revealed by use of the Flow Analysis system) passed back into and through the circle. The history of how this was done, and of the facilities which were obtained and later abused on the basis of the passage of this money through the circle of accounts was central to the fraud on the banks concerned, and was fully revealed by use of the analysis techniques described. A much simplified

Flow Analysis of one part of this process is used to illustrate this technique in Section 3.2.3 and **Figure 14** below).

### 17.3.2.3 Horizontal Tracing Flows of Money within a Circle of Accounts

Another whole area of work on cases involving transfers of money of great significance in both investigative work and in relation to the context of litigation is that of tracing flows of money horizontally through accounts.

Again, this is a wide subject, and the many purposes which it may have and the methods which may be used constitute a whole subject in themselves. I can describe here only very shortly two of the techniques used for horizontal tracing.

(i) Tracing and Quantification of Flows

It is often relevant to build up a picture of the general flow of funds through many accounts, and this can be done using the Link Analysis System briefly described under 17.2.2.4 above.

For this purpose, the links concerned are those arising from the payment of money from one party to another, or from one account to another. Using the search capacity of a data processing system, we can search for and analyse the chains of such links which arise from the transfers recorded in the database; and from this we can produce quantified representations of the flows of funds involved.

In performance of this task, the general data matching capacity of a data processing system are vital. In a large web of interconnected accounts, for instance, a search for all the chains of links between bank accounts which arise from transfers of money is likely to produce a completely unmanageable number of chains, possibly one or two thousand. We therefore use data matching techniques to refine the search. For instance, we may wish to include transfers only within a specified period, or only of a certain minimum amount, or only where the aggregate of transfers between two accounts within the period is of a certain minimum amount. We may also limit the links to those between certain parties, or parties with certain characteristics. The possibilities for refinement are very large. All the data for such refinements of course has to be in the data base. But if there, the computer can readily be instructed to check each possible link against the specified criteria and exclude those that do not meet them.

We have used the data base of the case which has already been described in Section 17.3.2.2 above to produce an example of such a Flow Analysis which is reproduced in **Figure 12**. For this purpose, we asked the system to trace money flow out from the party NM.AEA and, at the same time, to trace flows of shipments of goods back towards that party. In order to have a relatively simplified Link Flow

Chart for the purpose of reproduction, we refined the search by limiting it to flows of money in excess of an aggregate amount of $3,000,000, and to flows of goods in excess of an aggregate amount of $1,000,000, within the specified period.

While the actual Flow Analysis produced in the course of investigation of the account, which was less restrictively refined, was far more detailed and complex, the general picture of the transaction is clearly revealed in this simplified version. Thus, one sees the original flow of $42 million from NM.AEA to NM.AAD (the fraudster's name in which the many accounts in the consolidated circle were held) and the onward payments from NM.AAD of money to NM.AAA, BK.AER and NM.AEP. These 3 parties were all in one way or another linked to NM.AAD. From NM.AEP, funds flowed on to NM.AEW and NM.AEV. Both of these were genuine independent producers of the commodity allegedly to be shipped under the letter of credit issued by NM.AEA (see above 3.2.2 and **Figure 11**) The Flow Analysis also reveals that these last payments were fully covered by shipments of goods and that there was a substantial shipment of goods back from NM.AEP to NM.AAD. There appeared to be no covering shipment back from NM.AAD to NM.AEA. All of these aspects were the subject of further investigation using the same computer systems, and ultimately revealed the detailed mechanics of the transaction in ways which were critical, firstly, to the further outside investigation of the case, and, secondly, to proving allegations of fraud made in the course of litigation in the case.

(ii) Production of Flow Charts for Investigative and Presentational Purposes

One of the techniques used in further investigation of this transaction was the production of detailed Flow of Funds Analyses. As referred to above (17.3.2.2), it is impracticable to reproduce the Flow Analysis of the passage of the $42 million through the consolidated accounts due to its size and complexity. But we can illustrate this technique in relation to one small detail of the work involved in tracing the re-entry of some of the money back through the consolidated accounts.

This involves tracing, and presenting in comprehensible form for the purposes of evidence, the use made of some funds placed on deposit. This sum of $823,000 was placed on deposit in October, 1982 and designated as account AC.ABF. A study of the balances on the consolidated accounts (produced using another of the computer analysis systems) over the period, however, revealed that, though this deposit remained in place for a year, it did not result in any commensurate increase in the overall balance on the circle of accounts as a whole. We suspected that the money had been passed straight out of the circle back to the fraudsters; but bearing in mind that these balances were made up of some hundred accounts, many of them being extremely active current accounts, it was very difficult to see if any particular transfer out of the circle could, in fact, be proved to be directly linked to the deposit payment into it. We were able to do this, however, by use of the computer tracing techniques.    These produced the Flow Analysis which is reproduced in **Figure 13**.

This indicated that the sum originally deposited (designated account AC.ABF) was paid out to another account in the circle (AC.ADU, which was with a different bank) in October, 1983.    There, it covered an overdraft position which had been created by a transfer from AC.ADU to repay a loan which had been granted on the 1st December, 1982 on AC.AGO (which was a loan account with the same bank). The proceeds of this loan had been credited to AC.ADU and covered a debit position on that account resulting from a transfer of $816,837.75 to AC.ADR (a current account with yet another bank within the circle of consolidated accounts) from which $600,402.97 was paid out on the same date to NM.AFF (which was a company controlled by the fraudsters themselves).

### 17.3.3    Organisation and Presentation of Insurance/Re-insurance Data

It is important to emphasise the extent to which the same underlying techniques of data processing may be applied to a wide variety of situations.

This applies even to something as seemingly specific as the MFIND systems for organisation and analysis of money transfers.    For these systems can, in fact, be used as a basic analysis tool for any set of related transactions which involve what may usefully be defined as directional and quantifiable flows between units of some kind.    Besides transfers of money, transfers of goods also fall into this category, as do the issue of contingent liabilities such as guarantees or letters of credit.    There are many more.

A topical example at the present time can be taken from the field of insurance and re-insurance.    The issue of primary insurance risk can be seen as a quantifiable flow from underwriter to insured, as can each stage of the backward chain of re-insurance of that risk.    As in the case of transfers of money, a complex web of insurance risk builds up;    and this can be organised into notional "accounts" of which statements can be made up and which can be consolidated and analysed using the same routines as in the case of transfers of money;    and in the same way, flows of risk can be mapped out, quantified and analysed.

In fact, in the case of insurance risk accounts, as in the case of contingent liability risk accounts, we are likely to need to cross relate established risks to actual transfers of money which are made pursuant to and in discharge of the risk.    One of the ways of doing this would be to establish mixed accounts in which the establishment of risk would be recorded as a credit to the underwriter and debit to the insured, while payment under the insurance would be recorded as a flow in the opposite direction.

In the case of insurance risk, we would also be able to make analyses using general data matching techniques to relate the risk under a policy back to the individual underwriters in a syndicate, or shareholders in a re-insurance company.

### 17.3.4    Reconstruction Dealings on a Foreign Exchange Dealing Facility

Another example of the use of data processing techniques derives from a case involving the running by a bank of a foreign exchange dealing facility for a client.

This facility involved operation by the bank on behalf of the client of a complex system which required the build up and closure of numerous related long and short positions in a number of different currencies over a period.   We were concerned to investigate whether the bank had properly followed the agreed system, which involved clarification of the exact history of the build up and closure of each position, and to evaluate the net value of the client's total position with the bank on a daily basis (a value which was made up not only of the open positions involved, but also of a number of cash accounts, of sums placed from time to time on money market deposit and of interest and profit/loss accounts deriving from the dealing activities).

The bank had not produced any statements of the exchange dealing accounts, nor of separate interest or profit/loss accounts.   We had to work largely from the many hundreds of individual dealing slips, together with statements of the current cash accounts resulting from the dealing activities.

We used the system of consequential transactions at the input stage to reproduce the open position accounts (for which there were no statements) as well as the consequential cash accounts (for which there were, in fact, statements) direct from input of the data from the dealing slips.   This had to be supplemented with a very limited amount of additional data input to cover cash transactions, such as interest debits and credits and deposit placements, which did not derive directly from the dealing.

On the basis of this data, we were then able to analyse the numerous resulting open position accounts on the basis of each individual position, revealing the profit or loss resulting from its closure (which was then posted to a profit/loss account); and to produce complex daily balances which, firstly, revealed the way in which the bank had built up the required related long and short positions;   and, secondly, enabled us to value (at current exchange rates) the value on a daily basis of both the overall position of the client with the bank, as well as of a number of sub-divisions of that position which were relevant to complaints the client made against the bank.

**Figure 1**

**Text handling**

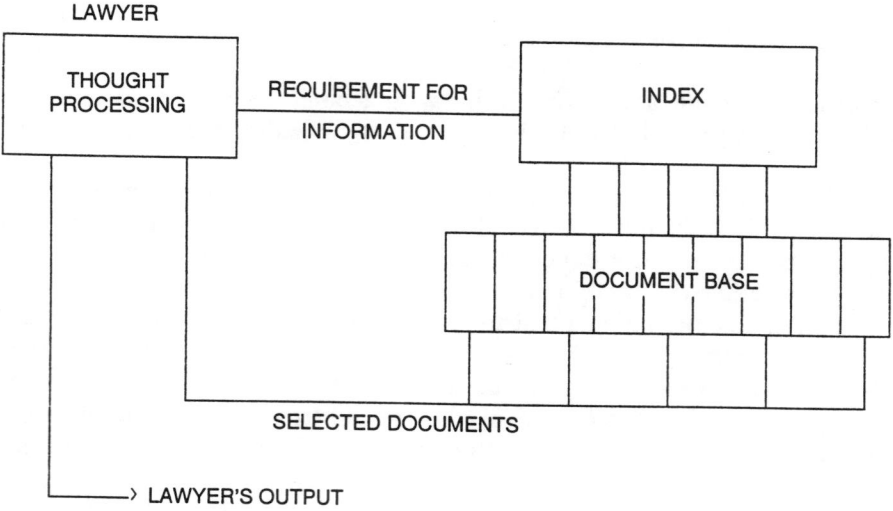

**Figure 2**

**Data Processing**

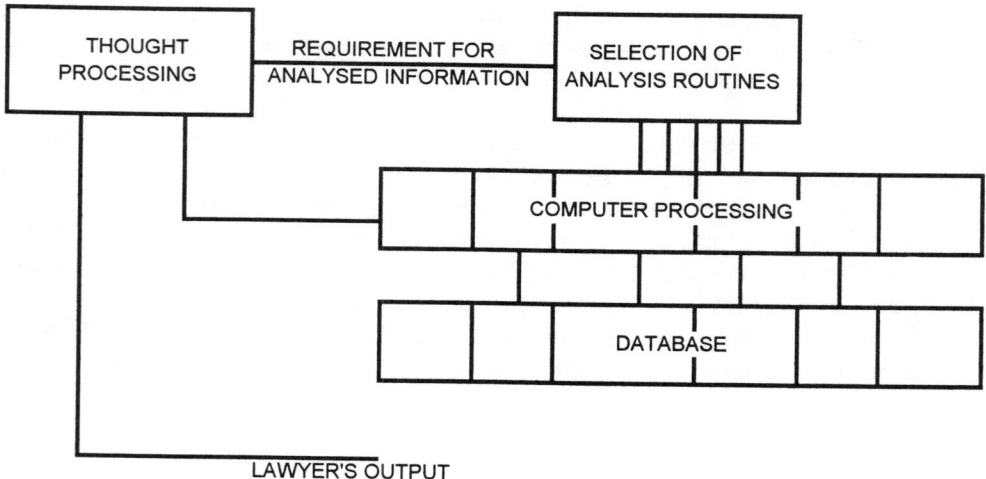

**Figure 3**

**3.1 Money Transfer Record**

| KEY | PAYOR | PAYEE | CURRENCY | AMOUNT | DATE |
|---|---|---|---|---|---|

**3.2 Money Transfer Records Grouped in File**

| | | | | | | |
|---|---|---|---|---|---|---|
| F | 1 | A | B | USD | 1,000 | 01.01.92 |
| I | | | | | | |
| L | 2 | A | C | GBP | 2,000 | 01.02.92 |
| E | | | | | | |
| | 3 | C | D | DMS | 5,000 | 01.03.92 |

**Figure 4**

**Relationship Between Names in Payor and Payee Fields of Money Transfer File and Data in Primary File for the Names**

MONEY TRANSFER FILE RECORD

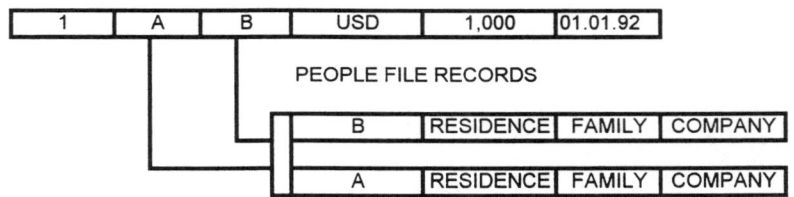

| 1 | A | B | USD | 1,000 | 01.01.92 |
|---|---|---|---|---|---|

PEOPLE FILE RECORDS

| B | RESIDENCE | FAMILY | COMPANY |
|---|---|---|---|

| A | RESIDENCE | FAMILY | COMPANY |
|---|---|---|---|

**Figure 5**

**Relationships between Cross Reference Index Records for Names and Transfer File Records in which those Names occur**

MONEY TRANSFER FILE INDEX

MONEY TRANSFER FILE RECORDS

| INDEX | | | | | |
|---|---|---|---|---|---|
| | 1 | A | B | USD | 1,000 |
| A | 2 | A | C | GBP | 2,000 |
| | | | | | |
| B | 1 | A | B | USD | 1,000 |
| | | | | | |
| | 2 | A | C | GBP | 2,000 |
| C | 3 | C | D | DMS | 5,000 |
| | | | | | |
| D | 3 | C | D | DMS | 5,000 |

204

**Figure 6**

Chain of links from index to money transfer file through Payee ("C") in record 2 in money file to "A" as director of Y Ltd, a Company in which "C" holds shares (names record for "C") and to "E" as owner of "100 x street, which is "C"'s Residence (Names record for "C").

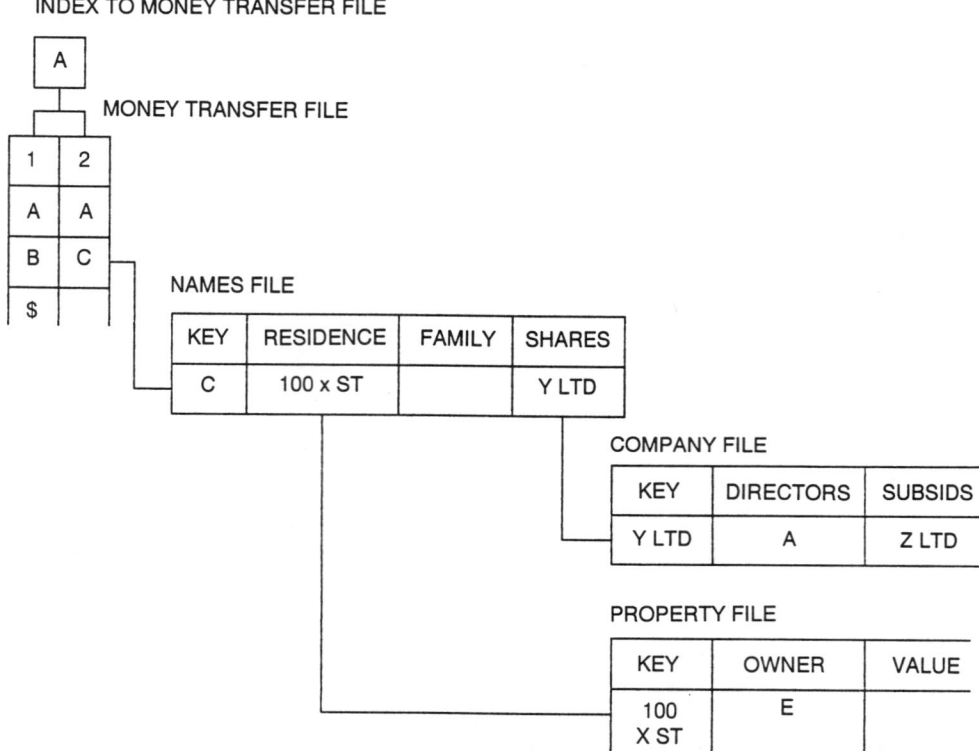

INDEX TO MONEY TRANSFER FILE

**Figure 7**

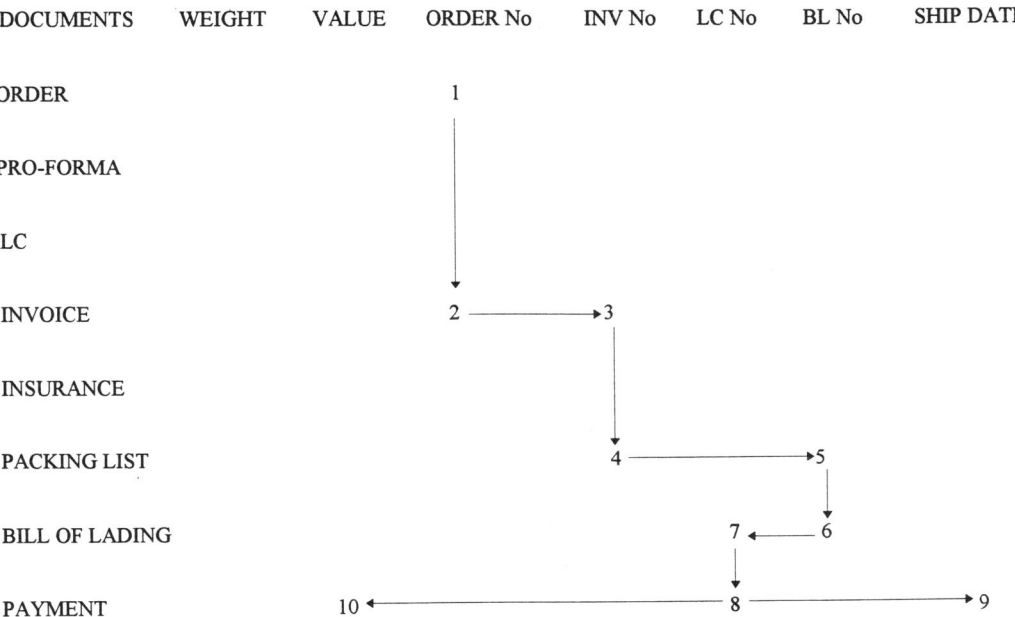

There were many shipments of same goods/weight value but Order no's were unique. An Order no. (1) reappeared on an Invoice (2) and the (unique) Invoice no. (3) appeared on a Packing List (4). The Packing List had a Bill of Lading no. (5) and the Bill of Lading had a LC no. (7). The LC no. identified several possible payments. But only one of these had both the right value (10) and a compatible shipping date (9). As a result of the links, we were able to group these documents as a set evidencing complete sale and shipment of goods.

206

**Figure 8**

| BANK A | | | | BANK B | | | | BANK C | | | |
|---|---|---|---|---|---|---|---|---|---|---|---|
| CRRNT | TRADE | DEP $ | DEP £ | CRRNT | TRADE | DEP $ | DEP £ | CRRNT | TRADE | DEP $ | DEP £ |
| CONS CREDIT | | CONS DEPO | | CONS CREDIT | | CONS DEPO | | CONS CREDIT | | CONS DEPO | |
| CONSOLIDATED BANK A | | | | CONSOLIDATED BANK B | | | | CONSOLIDATED BANK C | | | |
| CONSOLIDATED BANKS A + B + C | | | | | | | | | | | |

**Figure 9**

**MFM Services - Analysis of Movements on AC.AAA Cons 01.01.81 - 30.06.81 by Category ($US)**

| LIST | CREDIT | DEBIT | NETMOVES |
|---|---|---|---|
| Family | 0.00 | 8,039,664.14 | (8,039,664.14) |
| Associates | 0.00 | 49,497.14 | (49,497.14) |
| Partners | 244,698.21 | 1,986,937.87 | (1,742,239.66) |
| Family Group | 0.00 | 26,960,859.13 | (26,960,859.13) |
| IM Group | 0.00 | 99,926.59 | (99,926.59) |
| Miscellaneous | 31,751.06 | 377,144.73 | (345,393.67) |
| Consignees | 46,624,823.40 | 102,064.55 | 46,522,758.85 |
| Suppliers | 0.00 | 5,494,721.34 | (5,494,721.34) |
| Banks | 1,982,523.77 | 565,323.02 | 1,417,200.75 |
| **Total Selected** | 48,883,796.44 | 43,676,138.51 | 5,207,657.93 |
| Unknown | 4,117,902.97 | 4,127,533.61 | (9,630.64) |
| Unselected | 2,801.93 | 638,878.82 | (636.076.89) |
| **Total Unselected** | 4,120,704.90 | 4,766,412.43 | (645,707.53) |
| **GRAND TOTAL** | 53,004,501.34 | 48,442,550.94 | 4,561,950.40 |

**Figure 10**

**MFM Services - Payor Analysis for Period 01.01.81 - 30.06.81 ($US)**

| PAYOR | NUMBER OF PAYMENTS | TOTAL AMOUNT |
|---|---|---|
| NM.ABK | 7 | 2,203,430.85 |
| NM.AEA | 7 | 42,029,069.06 |
| NM.ACW | 2 | 54,310.78 |
| NM.ABX | 3 | 198,910.60 |
| NM.AEE | 3 | 200,657.56 |
| NM.AEG | 2 | 224,208.99 |
| NM.AEH | 1 | 22,325.06 |
| NM.AEI | 1 | 840,293.00 |
| NM.ADP | 1 | 4,249.38 |
| NM.AEJ | 2 | 156,134.44 |
| NM.AEK | 1 | 2,670.92 |
| NM.ACT | 1 | 387,133.58 |
| NM.ADO | 2 | 75,755.03 |
| NM.AEL | 1 | 52,154.00 |
| NM.AEM | 1 | 73,520.15 |
| **TOTALS** | 35 | 46,524,823.40 |

**Figure 11**

**MFM Services - Counter Party Analysis of AC.AEO**

| Date | Payor/Payee | Bank | A/C | Amount | | Balance | Bank Ref No | CONS AC | ID |
|---|---|---|---|---|---|---|---|---|---|
| 30.11.1980 | NM.AEA | | | (DHS) | $154,330.97 | | 3001/402125 | AC.ABR | 3030 |
| 10.01.1981 | NM.AEA | BK.ACH | | (DHS) | $5,944,338.76 | | 3001/402158 | AC.ABR | 3365 |
| 18.01.1981 | NM.AEA | BK.ACH | | (DHS) | $5,944,344.24 | | 3001/402158 | AC.ABR | 11006 |
| 21.01.1981 | NM.AEA | BK.ACH | | (DHS) | $7,935,721.92 | | 3001/402158 | AC.ABR | 11007 |
| 29.01.1981 | NM.AEA | BK.ACH | | (DHS) | $19,839,304.79 | | 3001/402158 | AC.ABR | 3477 |
| 10.02.1981 | NM.AEA | BK.ACH | | (GBP) | $925,670.58 | | 3001/101262 | AC.ABR | 3544 |
| 11.03.1981 | NM.AEA | | | (DHS) | $923,543.76 | | 3001/101260 | AC.ABR | 3670 |
| 22.03.1981 | NM.AEA | | | (DHS) | $516,145.02 | | 3001/101264 | AC.ABR | 3742 |

TOTAL CREDIT = 42,183,400.04       TOTAL DEBIT = 0       NET =       $42,183,400.04

**Figure 12**

**MFM Services - Link Flow Analysis from NM.AEA - MNY (F) ($3,000,000.00) : GDS    ($1,000,000.00)**
**Period 01.01.81 - 30.06.81**

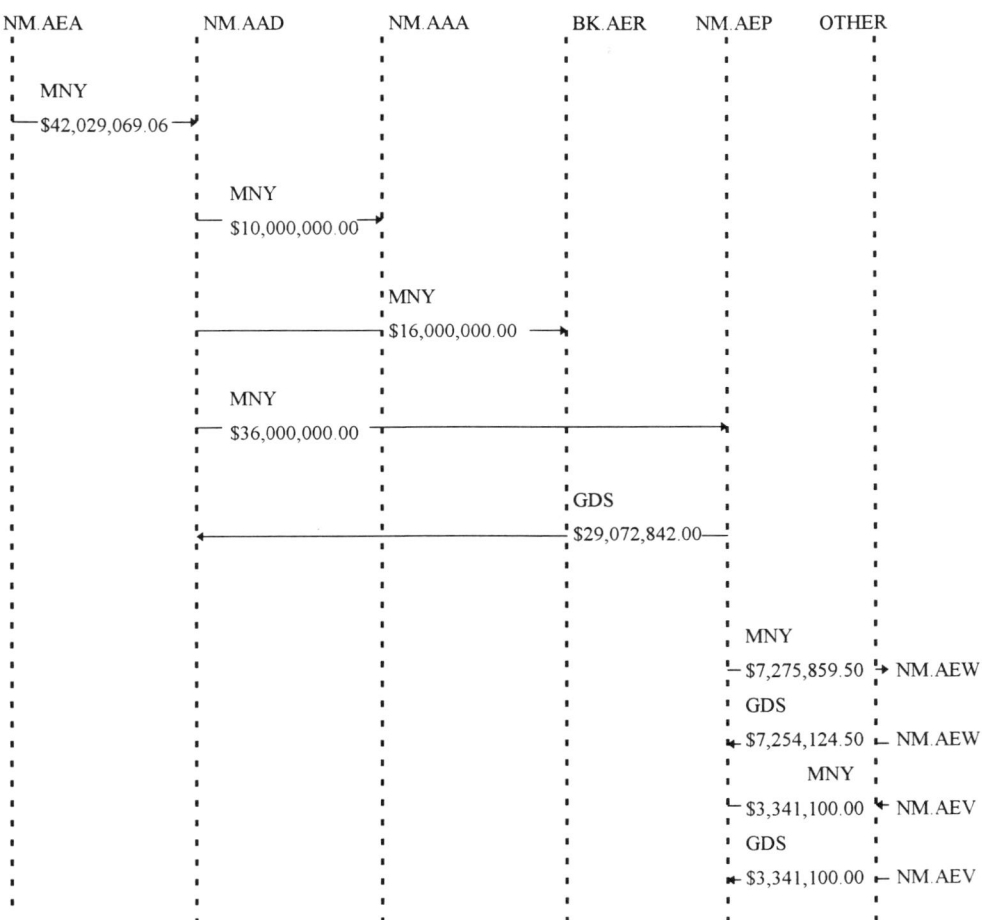

210

**Figure 13**

**MFM Services - Flow of Funds passed through AC.ABF to NM.AFF**

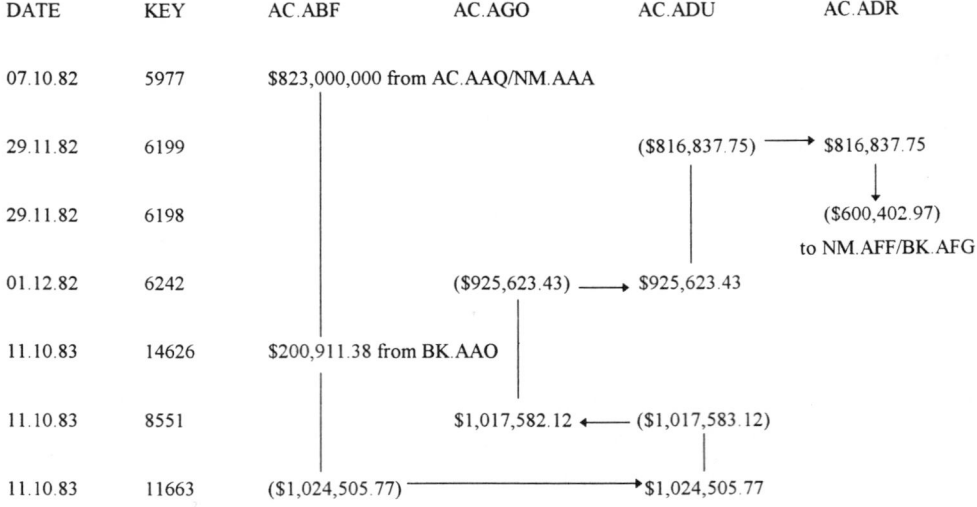

| DATE | KEY | AC.ABF | AC.AGO | AC.ADU | AC.ADR |
|------|-----|--------|--------|--------|--------|
| 07.10.82 | 5977 | $823,000,000 from AC.AAQ/NM.AAA | | | |
| 29.11.82 | 6199 | | | ($816,837.75) ⟶ | $816,837.75 |
| 29.11.82 | 6198 | | | | ($600,402.97) to NM.AFF/BK.AFG |
| 01.12.82 | 6242 | | ($925,623.43) ⟶ | $925,623.43 | |
| 11.10.83 | 14626 | $200,911.38 from BK.AAO | | | |
| 11.10.83 | 8551 | | $1,017,582.12 ⟵ | ($1,017,583.12) | |
| 11.10.83 | 11663 | ($1,024,505.77) ⟶ | | $1,024,505.77 | |

# CONTRIBUTORS

Dr V Mital
Canterbury Business School
University of Kent
Canterbury
Kent
CT2 7PE

R G Allison
Allison Consultancy
1 France Hill Drive
Camberley
Surrey
GU15 3QA

Mr Thomas J Heiden
Miller, Canfield, Paddock & Stone
1200 Campau Square Plaza
99 Munroe Avenue NW
P O Box 329
Grand Rapids
Michigan 49501
USA

Mr Richard Susskind
Masons
10 Aylesbury Street
London
EC1R 0ER

Ms Phyllis Deets
Gowen Deets
Litigation Management Services
286 Divisadero Street
San Francisco
California 94117
USA

Mr Joseph H Howie Jr
Docucon Inc
9700 IH-10W
Suite 200
San Antonio
Texas 78230
USA

*Phone 001 210 341 4188*
*Fax 210 342 1015*
*Curtis Hill Publishing Co*
*506 Sardan*
*Suite 150*
*San Antonio*
*Texas 78216*

Mr Graham Pearson
UK Quality Manager
Business Management Systems
        and Quality Division
Rank Xerox (UK) Ltd
Bridge House
Uxbridge, Middlesex

Mr John B Massopust
Larson & Zelle
33 South Street
City Center - Suite 4400
Minneapolis
Minnesota 55402
USA

Mr Peter Illion
Beachcroft Stanleys
20 Furnival Street
London
EC4A 1BN
Minnesota 55402
USA

Mr J D M Mackintosh
Lawrence Graham
190 Strand
London
WC2R 1JN

Mr Howard Field
Denton Hall, Burgin & Warrens
5 Chancery Lane
Clifford's Inn
London
EC4A 1BU

Ms Morag Macdonald
Bird & Bird
2 Gray's Inn Square
London
MC1R 5AF

Maurice E F Fitzmaurice
Maurice Fitzmaurice Services
3 Grays Inn Square
Holborn
London
WC1R 5AH

Ms Patricia S Eyres
Litigation Management & Training
        Services Inc
555 East Ocean Boulevard
Suite 700
Long Beach
California 90802
USA

*[handwritten: Phone 001 310 495 0098 1786]*

*[handwritten: 301 E Ocean Blvd Suite 220 Long Beach CA 90802]*

Mr Clive D Thorne
Denton Hall, Burgin & Warrens
5 Chancery Lane
Clifford's Inn
London
EC4A 1BU

Mr Neil Cameron
KPMG Management Consulting
P O Box 695
8 Salisbury Square
London
EC4Y 8BB

Ms Michele Gowen
Gowen Deets
Litigation Management Services
286 Divisadero Street
San Francisco
California 94117
USA

*[handwritten: 601 Gateway Blvd Suite 870 South San Francisco CA 94080]*

*[handwritten: 001 415 244 9600]*

*[handwritten: Fax 9719]*